# Calcium Hunger

The popular press recently has seized on the idea that calcium is important to human health. A feature article entitled "Calcium Takes Its Place as a Superstar of Nutrients" appeared in the *New York Times* in October 1998, and a week earlier, *Newsweek* ran a five-page advertising supplement, "Calcium Made Headlines This Year."

This book brings together the behavioral, physiological, and neuro-endocrine regulation of calcium and introduces an understanding of how the brain orchestrates whole-body demands for calcium. The approach is that behavior, in addition to physiology, serves bodily maintenance. The book links basic and clinical literature surrounding calcium homeostasis, as a wide variety of clinical syndromes are tied to calcium metabolism. Because calcium is so important during life stages particular to women, the book emphasizes the relevance of calcium to women's health, although this emphasis is not exclusive: calcium is fundamental to both sexes.

Jay Schulkin is a Research Professor at the Department of Physiology and Biophysics at the Georgetown University School of Medicine. He is also a research associate of the Clinical Neuroendocrinology Branch of the National Institute of Mental Health and Director of Research at the American College of Obstetricians and Gynecologists. He is the author of seven books, including *Sodium Hunger: The Search for a Salty Taste*, published by Cambridge University Press in 1991.

# Calcium Hunger

## BEHAVIORAL AND BIOLOGICAL REGULATION

---

**JAY SCHULKIN**

*Georgetown University School of Medicine*

CAMBRIDGE
UNIVERSITY PRESS

# CAMBRIDGE
## UNIVERSITY PRESS

32 Avenue of the Americas, New York NY 10013-2473, USA

Cambridge University Press is part of the University of Cambridge.

It furthers the University's mission by disseminating knowledge in the pursuit of education, learning and research at the highest international levels of excellence.

www.cambridge.org
Information on this title: www.cambridge.org/9780521795517

First published 2001

*A catalogue record for this publication is available from the British Library*

*Library of Congress Cataloguing in Publication data*
Schulkin, Jay.
Calcium hunger : behavioral and biological regulation / Jay Schulkin.
    p. cm.
Includes bibliographical references and index.
ISBN 0-521-79170-7 (hardback) – ISBN 0-521-79551-6 (pbk.)
1. Calcium – Metabolism – Regulation. 2. Calcium in human nutrition. 3. Appetite. I. Title.
QP535.C2 S33    2000
612.3'924 – dc21

00–034212

ISBN 978-0-521-79170-0 Hardback
ISBN 978-0-521-79551-7 Paperback

*This book is dedicated to Dick Denton and Paul Rozin.*

# Contents

# Preface and Acknowledgments

The origin of this book is my long-term interest in the regulation of minerals in the body. A primary research topic in my work has been sodium appetite. That research interest culminated in a book titled *Sodium Hunger: The Search for a Salty Taste* (Schulkin, 1991). Since then, my research interest in mineral homeostasis has shifted to calcium and the existence and mechanisms of calcium appetite. Much more is known about the mechanisms of sodium appetite than those of calcium appetite, but, if anything, that has only heightened my interest.

My excitement for this field can be traced back to the influences of a number of colleagues who have enriched my life. Many of them I have mentioned in previous books. Here I begin by mentioning two, and dedicating the book to them. They are Paul Rozin and Dick Denton.

I came to the University of Pennsylvania as a graduate student more than 20 years ago, in part because of Paul Rozin's research on thiamine deficiency. The experiments were simple and elegant; he did classic work on behavioral avoidance to diets deficient in thiamine. But he did a lot more than that. He was fun to be with, and he loved the culture of ideas. He could always engage. I can vividly remember running up the stairs to his office with my list of questions to ask him.

Dick Denton has always, to my mind, been the paragon of the statesman scientist: one who builds institutions (e.g., the Howard Florey Institute of Experimental Medicine at the University of Melbourne), who enjoys a rich life in which science is an entry to the larger world. Dick and I would sometimes meet at my apartment in New York City and take walks and talk. On the science side, I have always felt that no one is more knowledgeable or has thought more constructively than Dick Denton on the biological basis of sodium and phosphate hunger. Some 15 years ago in a café somewhere in New York City we talked about doing research on calcium appetite along the lines used for sodium appetite.

A conference on "Calcium Regulation over the Life Cycle" in 1998 helped to solidify and expand my thinking on calcium regulation. At that conference, Robert Heaney, working in the field for more than 30 years, always with a broad-based approach and unafraid, was a great inspiration for his knowledge and devotion to the task of understanding calcium regulation.

I thank Aviad Haramati, Kent Berridge, Heidi Kalkarf, Micah Leshem, Kristy Morris, Sue Mulroney, Ralph Norgren, Olav Oftedal, Jeff Rosen, Bert Slotnick, Alan Spector, and Mike Tordoff for their assistance and encouragement.

I want to thank, in particular, Mike Power. He works with me at the College, and I work with him over at the Smithsonian National Zoo. In both places, he is a great colleague.

Finally, my family and friends have been a great joy, and our new son Nicky has joined our family to enrich our lives.

# Introduction

The popular press in recent years has seized on the idea that calcium is important in human health. The title of a *New York Times* article in October of 1998 was "Calcium Takes Its Place as a Superstar of Nutrients." A week earlier, a five-page advertising supplement in *Newsweek* was titled "Calcium Made Headlines This Year."

Of course, following the recent rush to calcium frenzy, there likely will turn out to be disappointment. Calcium is no wonder cure nor panacea. It is, however, a basic element in the body. It is regulated in both behavioral and physiological terms. And it is important to ingest adequate amounts to maintain health.

There are not many books devoted to calcium regulation. One book, published some 40 years ago by a South African physician/scientist, suggested that "calcium is the most important inorganic element in the body" (Irving, 1957, p. 1). Perhaps he was overly enthusiastic about the importance of calcium. I do not know if it is the most important element, but to be sure it is critical. What Irving emphasized in his monograph was the absorption and utilization of calcium, the body's requirements for calcium, and the role of its excretion in maintaining calcium balance. Of course, he rarely mentioned behavior, let alone brain.

This book will provide a context in which the behavioral, hormonal, and physiological mechanisms for regulation of calcium can be somewhat understood. In addition, an understanding of how the brain orchestrates whole-body demands for calcium will be introduced. My approach is whole-body physiological regulation; behavior serves bodily maintenance in the same way that hormones and other physiological signals do.

The book also links the basic literature and the clinical literature. Several clinical syndromes are tied to calcium metabolism. The findings at the clinical level are less clear-cut and perhaps more controversial. I am not an epidemiologist, but I attempt to provide a format in which one can reason about the myriad clinical issues related to calcium homeostasis.

**Table I.1.** *Elements in Seawater*

| Element | Weight (%) |
|---|---|
| Cl | 58.2 |
| Na | 32.4 |
| Mg | 3.9 |
| S | 2.71 |
| Ca | 1.23 |
| K | 1.17 |

**Table I.2.** *Elements in the Human Body*

| Element | % of Total Number of Atoms | Weight (%) |
|---|---|---|
| H | 63.0 | 10.0 |
| O | 25.5 | 64.5 |
| C | 9.5 | 18.0 |
| N | 1.4 | 3.1 |
| Ca | 0.31 | 1.96 |
| P | 0.22 | 1.08 |
| Cl | 0.08 | 0.45 |
| K | 0.06 | 0.37 |
| S | 0.05 | 0.25 |
| Na | 0.03 | 0.11 |
| Mg | 0.01 | 0.04 |

## Basic Considerations in Calcium Regulation

In the ocean, excluding $H_2O$, which of course is not an element, calcium is the fifth most abundant element (Robertson, 1988). One can see that calcium is far less abundant than sodium (Table I.1).

Calcium is also the fifth most abundant element in the body (Table I.2). The concentration of calcium in the sea is higher than that in extracellular fluids (10 mmol/liter versus 2.5 mmol/liter). This is analogous to the modern sea having a higher concentration of sodium than do extracellular fluids (Dacke, 1979; Denton, 1982; Robertson, 1988). The elemental composition of the human body is shown in Table I.2 (Robertson, 1988). One can see that in vertebrates, the amount of calcium in the body is higher than that of sodium. That is entirely due to the skeleton, where 99% of calcium is stored.

The evolution of organisms from the high-calcium environment of the ocean to the low-calcium environment of dry land required physiological as well as morphological adaptations. But even before the colonization of land, the ability of sea-dwelling animals to store and utilize calcium had evolved. Land vertebrates further developed morphological and physiological systems to accumulate, store, and activate calcium ions when needed, while maintaining regulated levels of ionized calcium in extracellular fluids (Dacke, 1979; Nordin, 1988). The accumulation of calcium and phosphorus was essential for colonization of land by vertebrates.

There are common themes across species in the evolution of calcium regulation. But there are also important differences among species. In fish, for example, the gills, in addition to the kidney, play an important role in calcium regulation (Flik et al., 1995). Fish usually acquire calcium from the water through their gills (Flik et al., 1995). Active transport of calcium through membrane barriers is operative during development and periods of greater calcium need. Certain amphibians and reptiles have calcium-containing sacs that are utilized when there is calcium demand (Stiffler, 1996).

Calcium is regulated by tissue and also is a regulator of tissue in the body. Calcium must be derived from the environment and utilized in the body. The primary mechanisms that are involved are phylogenetically ancient and seem to be operative across species. Adaptations to excrete or retain calcium in crustaceans vary with the calcium medium in which they reside and depend on whether or not they are molting (Wheatley, 1996).

Calcium is a divalent metal that is fundamental for membranes and ionic regulation. Intracellular calcium is a "second messenger" and is mediated by cyclic adenosine monophosphate (cAMP). The range of biological functions that are regulated by intracellular calcium is quite diverse (Table I.3).

Intracellular concentrations of calcium are much lower than extracellular concentrations (Dacke, 1979; Robertson, 1988). The distribution of calcium in various tissues in a 70-kg man is depicted in Table I.4 (Robertson, 1988).

Numerous end organs participate in the regulation of calcium. The gastrointestinal tract and kidney, along with bone, are major regulators of calcium balance (Michell, 1986; Breslau, 1996a,b). Calcium receptors in the liver may monitor the calcium concentration in plasma, sending signals to the brain and affecting other peripheral sites in the body. These systems are carefully maintained via homeostatic mechanisms, and they underlie major functions such as blood coagulation, neuromuscular and neurotransmitter excitability, and enzymatic regulation.

The process of calcium regulation is dynamic. Of the approximately 1000 mg of calcium that should be ingested daily by a human, roughly one-third is absorbed by the intestines and then is transported into the

3

**Table I.3.** *Functions Regulated by Intracellular Calcium*

Excitation-contraction coupling in muscle
Neurotransmitter release
Cytoplasmic streaming
Control of cilia
Microtubule assembly
Membrane permeability to K and Ca
Exocrine- and endocrine-gland secretion of hormones
Egg fertilization
Cell division and reproduction
Cell-to-cell communication
Cyclic nucleotide metabolism
Certain enzyme activities (e.g., phosphorylase kinase)
Photoluminescence activity
Excitation of rods and cones
Chromosome movement
Initiation of DNA synthesis

**Table I.4.** *Distribution of Calcium in Human Tissue*

| Site | Calcium (g) | Weight (%) |
|------|-------------|------------|
| Skeleton | 1355 | 98.90 |
| Teeth | 7 | 0.51 |
| Soft tissue | 7 | 0.51 |
| Plasma | 0.35 | 0.026 |
| Extravascular fluid | 0.7 | 0.052 |

extracellular fluids. On a daily basis, up to 550 mg of calcium passes between bone and extracellular fluids (Breslau, 1996a).

Calcium is fundamental to physiological well-being, and there are hormones that act to regulate calcium in the body. Increased needs for calcium arise during certain periods of development: fetal and neonatal growth, pregnancy and lactation, and at the end of life.

There also appears to be an appetite for calcium, at least under certain circumstances. In that respect, calcium regulation appears to be another model incorporating behavioral and hormonal factors that affect internal physiological regulation. That is, the body may be "programmed" to ingest calcium when the need for more calcium arises.

**Figure I.1.** A pregnant woman and a lactating woman and the sorts of calcium-rich foods that they are likely to ingest. (J. Yansen and J. Schulkin, unpublished.)

Because calcium is so important during life stages that are particular to women (i.e., pregnancy, lactation, and postmenopausal aging), the relevance of calcium for women and for women's health is emphasized throughout this book. But obviously this is not completely a gender issue; calcium is fundamental for development and is needed throughout life for both sexes. However, calcium ingestion is altered during pregnancy and lactation, reflecting the greater metabolic needs during these periods (Laskey et al., 1998; Ritchie et al., 1998) (see Chapters 1 and 3). Figure I.1 depicts some of the rich sources of calcium that are ingested in general and perhaps are more necessary during pregnancy and lactation.

Although calcium regulation is a biological event, we are cultural animals. Our ingestive patterns reflect the cultural milieu. In fact, calcium ingestion worldwide is quite varied. A map depicting regional calcium ingestion is shown in Figure I.2 (Kanis, 1994). There are enormous variations across different cultures, and even within cultures. Even in the United States, intake is quite variable (Institute of Medicine, 1997).

A brief look at the foods that are advertised as rich in calcium will reveal the variety of potential sources of calcium (Figure I.3). Of course, salt

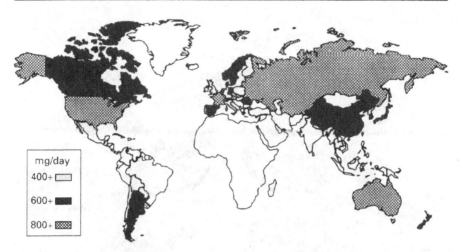

**Figure I.2.** Recommended dietary ingestions of calcium in different parts of the world. (From Kanis, 1994, with permission.)

licks are attractive to a number of species, though not often to humans. Mineral sources or salt licks can be found in dirt, stones, and mineral water (Figure I.4). Bones are also attractive to many animals and are rich sources of calcium, phosphate, and sodium; on bone appetite, see Denton (1982) and Cowan and Brink (1949).

## Structure of the Book

In each chapter there is a confluence of both animal and human experiments. The book begins with a discussion of the origins of the study of behavioral regulation of the internal milieu. For some investigators, the idea that behavior might have evolved to serve physiology still needs to be placed in perspective. Many of the behavioral concepts and their origins are given historical context and presented in light of calcium regulation (Chapter 1).

In Chapter 2 there is discussion in some detail of the behavioral regulation of calcium. Many of the behavioral concepts mentioned in Chapter 1 will find application in Chapter 2. For example, whether the appetite for calcium is innate or learned, the role of gustation in guiding behavior, and palatability judgments figure importantly in Chapter 2. In addition, the pica (e.g., chalk-eating) (Grigsby et al., 1999) behavior has been linked to alterations of calcium balance.

Chapter 3 returns to the theme of the evolutionary significance of calcium appetite. I argue that calcium appetite evolved in the context of the calcium

# Food Sources for Calcium

Listed below is a sampling of foods particularly rich in calcium.

| FOOD | QUANTITY | CALCIUM |
|------|----------|---------|
| Milk, nonfat | 1 cup | 300 mg. |
| Milk, Whole | 1 cup | 300 |
| Cheese, Swiss | 1 oz. | 260 |
| Cheese, American | 1 oz. | 200 |
| Ice cream, soft | 1 cup | 275 |
| Ice cream, hard | 1 cup | 200 |
| Shake, vanilla, thick | 11 oz. | 450 |
| Yogurt, whole 1 cup | 1 cup | 300 |
| Yogurt, low fat | 1 cup | 400 |
| Sardines, canned | 3 oz. | 375 |
| Oysters, raw | 1 cup | 225 |
| Salmon, pink, canned | 3 oz. | 150 |
| Black-eyed peas | 1cup | 200 |
| Tofu (soybean curd) | 4 oz. | 150 |
| Whole sesame seeds | 3 Tbsp. | 300 |
| Broccoli, cooked | 1 stalk | 150 |
| Bok choy, cooked | 1 cup | 250 |
| Collards, cooked | 1 cup | 300 |
| Okra, cooked | 1 cup | 150 |
| Spinach | 1 cup | 200 |
| Apricots, dried | 1 cup | 100 |
| Raisins | 1/3 cup | 30 |
| Orange | 1 medium | 50 |
| Rhubarb, cooked | 1 cup | 212 |
| Molasses | 1 Tbsp. | 150 |
| Sugar, brown | 1 cup | 175 |
| Almonds | 1/2 cup | 150 |
| Peanuts | 1 cup | 100 |
| Walnuts | 1/2 cup | 150 |
| Mineral water, Perrier | 1 liter | 140 |
| Mineral water, Ferrarelle | 1 liter | 446 |
| Mineral water, Mendocino | 1 liter | 380 |

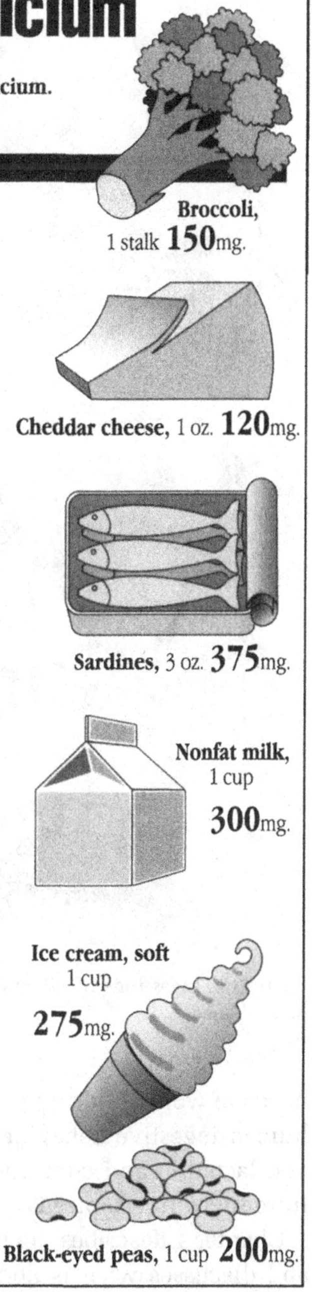

Broccoli, 1 stalk **150**mg.

Cheddar cheese, 1 oz. **120**mg.

Sardines, 3 oz. **375**mg.

Nonfat milk, 1 cup **300**mg.

Ice cream, soft 1 cup **275**mg.

Black-eyed peas, 1 cup **200**mg.

**Figure I.3.** Food sources rich in calcium. (J. Yansen and J. Schulkin, unpublished data.)

**Figure I.4.** Ingestions of mineral sources by different species. (J. Yansen and J. Schulkin, unpublished.)

needs of women during pregnancy and lactation. I discuss the literature on human ingestive behavior and physiological regulation during pregnancy and lactation and continue the discussion on pica behavior that was first introduced in Chapter 2.

Chapter 4 describes the neural structures that underlie ingestive behavior and discusses what is known about the roles of various steroid and peptide hormones that may influence calcium ingestion by their actions in the brain, by altering calcium-binding proteins, calcium transport mechanisms, or calcium receptors. The chapter ends with a framework in which to understand regions of the brain that may underlie calcium ingestion.

Chapter 5 discusses the nutritional and clinical issues and some basic science that underlies calcium metabolism. This chapter reiterates some of the themes of the earlier chapters (e.g., bone metabolism, gonadal hormones and calcium regulatory hormones, hypocalcemia and hypercalcemia) in the context of describing calcium ingestion over the life cycle. I then discuss some of the important roles of calcium in health and disease states that have been linked to calcium metabolism.

The Conclusion returns to the general theme of calcium regulation and how the study of calcium metabolism is amenable to behavioral, clinical, physiological, and molecular levels of analysis. But we are cultural animals, and I return to consider whole-body physiological regulation within the cultural milieu of food choice.

## CHAPTER ONE

# Behavioral Regulation of the Internal Milieu

There is a long tradition of investigation – harking back to Claude Bernard and Walter Cannon and further advanced by Curt Richter – that has demonstrated how behavior and physiology often serve the same end of maintaining homeostasis. For example, when extracellular fluid and sodium are depleted in the body, the concentrations of hormones that act to conserve sodium become elevated, in addition to their action in the brain to generate a hunger for salt (Denton, 1982; Epstein, 1982; Fitzsimons, 1998). Thus behavior and physiology are serving the same end, namely, to maintain homeostatic balance. Evolution has selected in favor of a variety of both behavioral and physiological conservation and excretion mechanisms to regulate the internal milieu. This apparently holds true for calcium regulation.

Four principal figures laid the biological groundwork for an understanding of the maintenance of calcium balance: Charles Darwin, Claude Bernard, Walter Cannon, and Curt Richter.

## Adaptation

From Darwin's work in the nineteenth century, two principles emerged that changed the landscape of biological study (Darwin, 1958, 1965). The first was the idea of evolution by natural selection. While we take that concept for granted now, just try to imagine what it must have meant at the time! The second was the recognition of the evolution of sexual characteristics. Both involve the concept of adaptation.

Adaptation is the key to biological evolution. Adaptation is also the key to the design of individual organs that regulate the internal milieu. Organs serve functions that are the results of natural selection. Kidneys evolved to regulate, in part, fluid balance. It is not that kidneys are ideal or perfect systems. They are not. An example of natural selection in animals can be seen among marine birds. The bird kidney is unsuited to ingest saline water, but marine birds have an accessory gland (known as the nasal or salt filter) that excretes excess sodium from food or ingested seawater (Schmidt-Nielson, 1997).

Adaptationist arguments about organs or behaviors have limits (Gould, 1977). But adaptation is one of the principal biological concepts, and it is fundamental in reasoning about the regulation of the internal environment. It is also basic in understanding the way in which the external environment is utilized to maintain the internal milieu.

Sexual dimorphic characteristics typically are features such as hair, horns, and genitalia. These are the common parlance of sexual dimorphism. Genders, like species, are distinguished on the basis of morphological features. But sexual dimorphic characteristics might be extended to include physiological and behavioral features, such as the need for and ingestion of calcium during reproduction in some animals. Other examples of such physiological sexual dimorphic features include the changes in vitamin D levels that occur in females during pregnancy to help maintain calcium, as well as variations in other hormone levels in specific circumstances (see Chapter 5).

Other concepts that were fundamental to Darwin's theories were variation and speciation (the evolutionary process by which new species are formed). Variation will play a part later in this book in some of the discussions of behavioral and physiological regulation of calcium. Moreover, what Darwin helped to solidify was an appreciation of the vast number of species and their evolution and the myriad ways of adapting to the problems of surviving and reproducing (Mayr, 1982, 1992). This is perhaps best seen in his diaries and his descriptions of many species in South America (e.g., the iguana, *Iguana iguana*).

Coincidentally, that species has also been studied in regard to its calcium regulation and the role of vitamin D (Oftedal et al., 1997). Iguanas in zoos are vulnerable to bone demineralization, coinciding with extremely low levels of vitamin D in the systemic circulation (Allen et al., 1995). One result is calcium depletion from bone. We will see in a later chapter that ultraviolet light plays an important role in the genesis and maintenance of vitamin D levels (DeLuca, 1988; Holick, 1994). Perhaps levels of vitamin D are important in the behavioral regulation, as well as the physiological regulation, of calcium (Oftedal et al., 1997).

In one study, iguanas were offered a calcium source (finely ground oyster shell in a bowl) in addition to their normal food and water. The bowl placement was alternated to avoid what psychologists call "place preference." The bowl was weighed daily. The calcium source was ingested in significantly higher quantities by the iguanas with higher levels of vitamin D than by those with low levels (Oftedal et al., 1997). That was an example of behavioral regulation of a bodily need, as depicted in Figure 1.1.

Returning to Darwin, his original theories have had to be adapted somewhat over the past 150 years (e.g., Gould and Eldridge, 1977), but the key

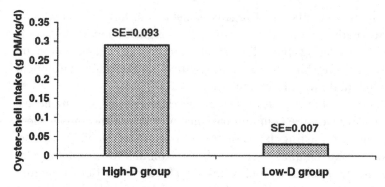

**Figure 1.1.** Oyster-shell ingestion by iguanas with high levels of vitamin D and low levels of vitamin D. (Adapted from Oftedal et al., 1997.)

concepts remain, and they figure importantly in understanding calcium regulation, as we will see in later chapters: Namely, behavioral and physiological adaptations are selected by evolution.

## Internal Milieu

Claude Bernard's 1865 treatise, *An Introduction to the Study of Experimental Medicine*, was a valiant attempt to render biological knowledge and the process of scientific discovery understandable to the general public. He lamented that scientific research was not presented in terms that were intelligible for a wide audience, and in that treatise he aimed to make the world of scientific discovery part of the parlance of educated people.

That treatise is a study in the philosophy of the experimental method in medicine and in the introduction of the idea of mechanisms within biological systems. For Bernard, "a living organism is nothing but a wonderful machine endowed with the most marvelous properties and set going by means of the most complex and delicate mechanism" ( Bernard, 1957, p. 63). Since Bernard's time, the notion of a "machine" that functions according to strict clockwork and out of necessity has been reworked to encompass more sophisticated concepts, including the idea that there are mechanisms that underlie flexibility, choice, and variability.[1] Bernard's work on glucose regulation, though not directly relevant to calcium regulation, was important for envisioning the means for regulation of biological systems. Most important, he introduced the concept of the "internal milieu" (Bernard, 1957). He concluded that in the internal environment, the transport of glucose in and out of cells was carefully regulated. That was an important idea – that internal organs were defended, that systems engaged in active self-preservation.

## Homeostasis

How the body regulates the internal milieu was the subject of study for twentieth-century physiologist Walter Cannon (1929).[2] Cannon's work was oriented toward homeostatic regulation, a concept he introduced and made part of scientific parlance (Cannon, 1966; see also Starling, 1923). Organs had evolved to maintain themselves within a largely constant environment. Glucose and sodium, for example, had to be maintained within strict limits by physiological means. Homeostasis was a way to keep the body in balance.

Cannon tried to understand how demanding situations affect normal physiological processes. For example, what happens to the body's demand for glucose when one is afraid? In such a situation, there are greater metabolic demands; cortisol is secreted, and energy is used, as Cannon discovered (e.g., Dallman et al., 1995).

Interestingly, Cannon singled out calcium in his 1932 book, *The Wisdom of the Body* (Cannon, 1966). He noted how calcium was stored, its distribution in the body, its maintenance by hormones such as vitamin D and parathyroid hormone, and its link to diseases like tetany. Moreover, he cited the increased demands for calcium at certain times, especially by pregnant women and women nursing infants. He noted that "although calcium is obviously needed by all of us in continuous excess during the period of life when the bony framework of the body is being formed, there are times in the experience of a woman when the demand for calcium is especially great. During pregnancy she must provide calcium for the developing fetus, and throughout the months of nursing she must provide an even greater amount in the milk, in order to give the baby the calcium requisite for growth" (1966, p. 140).

Cannon also studied the role of calcium in the body and showed the effects of low- and high-calcium diets on the bones of cats. He demonstrated that cats on a calcium-deficient diet had decreased amounts of trabeculae in the humeri, suggesting that the body stores calcium in the trabeculae and depletes those stores during times of calcium deficiency (Figure 1.2).

## Behavioral Regulation

Curt Richter was the originator and foremost exponent of the central concept that behavior serves the same end as physiology. He has been characterized (I believe rightly) as a "complete psychobiologist" (Rozin, 1976c).[3] Richter was profoundly influenced by Cannon's dictum in *The Wisdom of the Body*: The body "knows" what is best; it has ancient biological knowledge embodied in its tissues. For example, in contexts where there was a choice of

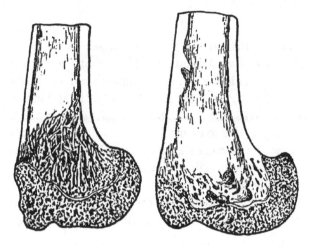

**Figure 1.2.** Sketches (following a photograph) showing the effects of diet on the trabeculae in the humeri of a cat. The humerus on the left, removed after a high-calcium diet, contains many more trabeculae than does its opposite member, which was taken after a low-calcium diet. (From *The Wisdom of the Body*, Revised Edition by Walter B. Cannon, M.D. Copyright 1932, 1939 by Walter B. Cannon, renewed © 1960, 1967, 1968 by Cornelia J. Cannon. Used by permission of W.W. Norton & Company, Inc.)

nutrients, that knowledge was embodied in the brains of animals. Bernard emphasized the regulation of the internal milieu, and Cannon the concept of homeostatic regulation, both of which apply to physiological regulation. What Richter introduced and demonstrated was the idea that behaviors had evolved to serve physiological regulation – that the same mechanisms that might physiologically promote homeostatic balance would also generate behaviors that would fulfill the same end (1943, 1956).

Richter was a whole-organism physiologist – a rare breed at the time, and even more rare today. Nutritionists such as E. V. McCollum at Johns Hopkins University had a significant influence on Richter regarding how variations in diet can influence behavior. For example, a low-sodium diet results in greater sodium ingestion, a low-protein diet in greater protein ingestion, a low-vitamin diet in greater vitamin ingestion. Richter, perhaps mistakenly, proposed a whole slew of what he thought were innate behavioral tendencies linked to dietary choice.

In his nearly 60 years working at the Johns Hopkins University Medical School, Richter often linked behavioral and physiological adaptations to medical issues. For example, when a young boy was admitted to Hopkins, his mother reported that he was eating excessive amounts of table salt (Wilkins and Richter, 1940). He was placed on a low-sodium diet. He died. When he was autopsied, it was noted that he had a pathological adrenal condition.

He had not been synthesizing the sodium-retaining hormone aldosterone, and thus he had chronically been losing sodium. His behavior of sodium ingestion had been an adaptive response, the same as that observed by Richter (1936) in adrenalectomized rats. His sodium ingestion had been a necessary behavior for survival that tragically had not been recognized by the attending physicians.

Concurrent with Richter's work were the self-selection studies that Clara Davis (1928, 1939) carried out at Children's Memorial Hospital in Chicago. She tested whether or not neonates would select a nutritional diet from a range of choices. The food sources that she offered them were wide-ranging and nutritionally wholesome, and the study continued for a six-year period. Such a study would be unimaginable today. Nonetheless, Davis's results suggested that when neonates were presented with a wide array of nutritional sources, they would select a balanced diet.

Many researchers have doubted the validity of the self-selection theory, questioning whether or not much of what Davis and Richter had observed had been artifactual, and suggesting that she had biased the ability of the children to choose appropriate sources by her selection of the types of foods presented (all whole foods were dense in nutrients) and the way in which they were presented (Rozin and Schulkin, 1990; Galef, 1991, 1996). Davis's intellectual milieu was replete with the concept of animals' ability to self-select a nutritional diet.

Richter (1943) conducted his experiments in self-selection using rats. Typically he offered rats a choice among a wide range of nutrients and minerals that are essential for health. He found, as had Davis in neonates, that rats would self-select what they needed to survive, that they could select an adequate diet. In one study, Richter demonstrated that rats that self-selected their food from a range of choices experienced growth, reproduction, activity, and overall health similar in nearly all respects to those for rats that were given a controlled diet of sufficient nutrients. The results were construed as behavioral homeostasis writ large.

Richter of course looked at calcium appetite. In several studies he used a parathyroidectomized-animal model. When the parathyroid glands are removed, parathyroid hormone, which is fundamental for conserving calcium, is not produced, calcium is poorly absorbed, and calcium is excreted in the urine. The animals start to lose calcium via the urine. Tetany is one outcome of this condition. He wanted to see if such animals would choose to ingest calcium. After all, their survival was at stake. He found that parathyroidectomized rats and monkeys would indeed increase their voluntary calcium ingestion (e.g., Richter and Eckert, 1938a; Richter, 1965). This phenomenon will be discussed in more detail in Chapter 2. For now, note the behavioral

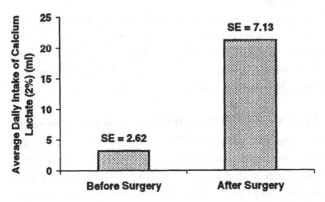

**Figure 1.3.** Ingestion of calcium by parathyroidectomized rats over 24 hours. (Adapted from Richter, 1955.)

change in ingestion of calcium as a result of parathyroidectomy, shown in Figure 1.3.

## Nutritional Choice

Of course Richter tended to be a nativist, that is, he believed that appetites were innate (Katz, 1937; Richter, 1956). Others argue that appetite relies on experience, on learning the beneficial effects of ingestion and amelioration. Harris et al. (1933) studied vitamin B appetite, where there is clearly a learned component. Nutritional choice involves both innate predilection and learning. The interesting question is how nutritional learning experiences interact with innate knowledge. It is not a question of one or the other, isolated from one another. It is a whole-body affair: innate mechanisms, the recognition of a particular taste, the detection of a tissue deficit, and amelioration of the tissue need.

One contemporary investigator has made a point of providing a "contrarian" view (Galef, 1991) of the self-selection experiments. Galef suggests that, except for the cases of sodium and phosphorus, where he concedes that the evidence has been good, the extent to which there are specific hungers for other nutrients is questionable (cf. Kon, 1931; Lat, 1967; Collier et al., 1969a,b). Galef, however, never mentions calcium. Galef criticizes the self-selection paradigm as an oversimplification of behavior: Not every appetite is the result of an innate need to provide the body with a specific substance. And this contrarian view has the ring of sanity amidst the exaggerated claims of encompassing behavioral regulatory competence. Perhaps the studies that focused on the selection of nutrients in ideal and artificial situations and the controversy they generated led many scientifically minded nutritionists to eschew behavioral considerations.

There are other theories of ingestive behavior that complement the principle of internal regulation, such as the concept of avoidance. Studies have demonstrated the ability of animals to avoid foods that would make them ill (later termed "taste-aversion learning") (Rozin, 1976a,b). For example, one study evaluated thiamine deprivation in laboratory rats (Rozin, 1967a,b). Thiamine deficiency in humans (beriberi) results in anorexia and decreased body weight. Similar reactions occurred when thiamine was eliminated from the diet of weanling rats. In response to the effects of thiamine deprivation, the rats voluntarily switched from their thiamine-deficient diet to an alternative diet. The rats would ingest the new diet whether it had thiamine in it or not.

The process of learning that a new diet was safe was not immediate; in fact, such learning was considerably delayed (Garcia et al., 1955, 1974; Rozin, 1976b). Assessment of the relationship between ingestion and the restoration thiamine of balance took hours. One adaptive strategy used by animals in the laboratory was to sample a source and then wait to determine the consequences of the ingestion (Rozin, 1976a,b).

Of course, which strategy an animal can use reflects its evolutionary constraints. We (like rats) are omnivores. We can switch readily, though warily, to a new food source. The koala, a true food specialist, is constrained to a very limited food repertoire (Rozin, 1976a).

There is, however, some evidence of specific hunger for proteins and carbohydrates (Osborne and Mendel, 1918a). Dilution of dietary protein results in preferential ingestion of a protein diet (Rozin, 1968). There is some evidence of immediate ingestion of protein among protein-deficient rats (e.g., Booth and Simpson, 1971; Anderson, 1979; Deutsch et al., 1989), but that that reflects innate specific recognition is highly questionable (Rozin and Schulkin, 1990). But that behavior appears to be influenced by the palatability of the diet; the less palatable the diet, the less probable it is that the rats will ingest the source of protein (Beck and Galef, 1989; Galef, 1999).

Rats placed on a carbohydrate-deficient diet will preferentially ingest carbohydrates when offered (e.g., Scalafani, 1987; Leibowitz, 1995). There may also be a strong innate bias to ingest sweet substances, as simple sugars are quick energizers, but sweet substances may not be helpful for long-term energy regulation. Long-term regulation requires protein ingestion (Dallman et al., 1995).

In summary, though there have been numerous experiments that have demonstrated that animals in a wide variety of self-selection paradigms can regulate a number of metabolic, caloric, fat, protein, mineral, and water balances (Richter, 1943), good sense dictates that nature did not confer specific hungers for so many nutrients and minerals. Even the great omnivores could not select a balanced diet in self-selection experiments on the basis of dietary categories alone (Rozin and Schulkin, 1990).

## Innate and Learned Behavioral Categories

It is important to distinguish between innate behavior and learned behavior – that is, what underlies behavioral choice. Behavior characterized as innate typically does not require prior experience for an immediate recognition and response (Chomsky, 1972; Pinker, 1997). By contrast, learned behavior requires, at a minimum, associative mechanisms (Rescorla, 1981). Both can be linked to innate qualities, up to a point. For example, there is innate aversion to bitter tastes, and innate attraction to sweet tastes. Sodium and calcium are in the mixed category; they are experienced as neither solely attractive nor aversive (Berridge et al., 1984; Bartoshuk, 1988, 1991; Tordoff, 1996b).

The foregoing example of thiamine deficiency is a classic example of a learned appetite. The appetite for sodium (as discussed later) is a classic example of an innate specific appetite. Sodium, in fact, may be the only mineral to elicit such a clearly specific appetite (Wolf, 1969a; Denton, 1982). No other mineral has so distinctive a taste as NaCl (Bartoshuk, 1974; Schulkin, 1991c). The question is, What is a specific hunger? Can it be narrowly defined as a hunger that automatically prompts an animal to realize when it is deficient in a certain mineral and then triggers the appropriate sensory/cognitive apparatus to detect and ingest sources of that mineral from the environment? How specific is "specific"?

In the context of innate and learned behaviors, it is important to ask these questions: What is the specificity of the response? How might the response have evolved? What natural selective pressures are there for the behavioral response? It is also important to take into account animals' learning strategies, such as neophobia (avoidance of new things) and latent learning (recognizing that a resource is located at a given site, or recognizing its association, and then being able to return to that resource when it is needed) (Tolman, 1949; Krieckhaus, 1970; Paulus et al., 1984).

I will suggest in what follows that behavioral regulation of calcium does not fit into either of these categories as nicely as does sodium appetite (the paradigmatic example of an innate appetite) or thiamine deficiency (the paradigmatic example of a learned appetite). Calcium appetite has features of both. As we will see in later chapters, there is a rich behavioral regulatory system that underlies calcium ingestion in animal models. But pica, the irrational ingestion of unusual substances, reveals itself in the calcium literature (Wood-Gush and Kare, 1966; Snowdon, 1977).

## Palatability Judgments

Kent Berridge and his colleagues (Berridge and Grill, 1984; Berridge and Schulkin, 1989) have used the term "palatability judgment" to describe a specific reaction to a taste. They point out that the palatability judgment of a substance is not the same as the sensory characteristics of that substance.

Underlying behaviors there are broad-based approach/avoidance mechanisms (Schnierla, 1959; Stellar and Stellar, 1985) that are linked to sensory stimulation. The more intense the stimulation, in a number of instances, the greater the avoidance behavior (Schnierla, 1959).

The motivation for ingestion of a mineral can be characterized as having at least two phases: search and ingestion (Craig, 1918; Schulkin et al., 1985). But, as discussed later, ingestion is not the same as hedonic satisfaction (Berridge, 1999). The search is known as appetitive behavior. The ingestion is consummatory behavior. Animals that forage for food obviously exhibit both. Both behaviors require an animal to use sensory detectors and physiological regulatory systems to maintain the behavioral response.

Some years ago, Michel Cabanac (1971, 1979) coined the term "allesthesia" to describe the fact that attractions toward some substances (such as the perceived taste or temperature of a substance) (Stellar, 1974, 1982) can change with the body's changing needs for the nutrients that those substances contain. Allesthesia is clearly reflected in the oral and facial responses reported in both animal and human studies. For example, rats that are sodium-deficient find hypertonic NaCl solutions (seawater concentrations) highly palatable, whereas rats with adequate sodium levels will reject seawater (Berridge et al., 1984; Flynn et al., 1992). In such a case, homeostatic regulation is linked to behavioral and physiological regulation of the internal milieu.

The facial expressions shown by human infants in response to various tastes are similar to those of other primates (Steiner et al., in press). In the case of neonates, infusions of sucrose or sweet-tasting substances will elicit positive facial responses; infusions of quinine will do the converse (Steiner, 1979). These responses are brain-stem-mediated (Grill and Norgren, 1978): Anencephalic children demonstrate the same responses to sweet and bitter ingesta. The authors of a study comparing human-infant responses with those of 11 other species concluded that the human and primate facial responses were in fact indicators of pleasure or displeasure, of positive or negative emotions elicited by taste, not simply responses to taste alone (Steiner et al., in press).

Some stimuli are naturally attractive or alluring. In fact, it was Charles Darwin who highlighted the fact that hedonic experiences underlie our reactions to food sources. His 1872 book *The Expression of the Emotions in Man and Animals* is rich with examples of alimentary reactions to foods (Darwin, 1965). Such innate hedonic judgments help to influence ingestion and may reflect the fact that foods that are normally nutritious in nature are palatable, and those that may be harmful tend to taste bitter (Pfaffmann, 1967; Janzen, 1977). Nature links the regulation of the internal milieu, in part, to judgments about the things that taste good and bad.

Calcium has a complex combination of tastes, unlike salt/sodium substances or sweet substances, which have simple and distinctive tastes. As with sodium, there appears to be some preference for dilute calcium solutions among calcium-replete rats (Tordoff, 1994).

On the other hand, the hedonic response does not always (and, in many instances in some cultures, only rarely) serve the nutritional needs of the body (Young and Chaplin, 1949; Galef, 1996). It is, of course, possible to enjoy eating substances that one does not need to eat (such as Good & Plentys). That should not be news to any of us humans. The fact that we eat a lot of "junk" does not mean that some aspects of our ingestive behavior do not reflect metabolic or mineral requirements under some conditions.

But there is a distinction between liking and wanting (Berridge, 1996, 1999). "Liking" relates to palatability or pleasure, whereas "wanting" relates to appetite or incentive. A person who has not eaten for many hours is hungry because the body requires food; this is an example of wanting. A person who has eaten a satisfying meal and then eats dessert may be eating for the pleasure of the taste of the dessert; this is an example of liking. A change in intake can be related to a change in liking or a change in wanting.

## Behavioral Anticipation of Nutritional Needs

There is another behavioral observation that has relevance to ingestive behavior and to behavioral events in general. It emerges from the literature on how biological clocks (e.g., circadian, semiseasonal) organize behavioral and physiological events that serve an animal (Richter, 1965; Rosenwasser and Adler, 1986). The seasonal secretion of the hormones of reproduction in seasonal-breeding animals is but one example in which internal clocks regulate physiological responses and subsequently behavioral responses that serve the animals.

With regard to regulation of the internal milieu, there is a distinction between reactive homeostasis and predictive homeostasis (Moore-Ede et al., 1982; Moore-Ede, 1986). Reactive homeostatic responses are the corrective measures taken after some imbalance, such as a decrease in extracellular fluid volume after excessive sweating, or loss of blood. The result is activation of the hormones of sodium balance to conserve sodium. That situation might also elicit the behavior of sodium ingestion to correct the imbalances (Denton, 1982). Anticipatory homeostatic responses are also hallmarks of ingestive behavior, such as water regulation (Fitzsimons, 1979).

Predictive homeostasis – what an animal might need – operates in anticipation of the tissue deficit or of greater need. Perhaps that is why female animals go to salt licks when they are not sodium-deficient – they are anticipating the sodium needs of pregnancy and lactation (Denton, 1982; Schulkin, 1991a,c).

As a feature of evolution, organisms move from passive responses to metabolic changes to anticipatory responses. A concept that entails some of the anticipatory aspects in behavioral events is allostasis. This term emerged in the scientific literature to account for both behavioral and physiological events that were not strictly homeostatic (Sterling and Eyer, 1988; Schulkin et al., 1994a; McEwen, 1998).

## Ecological Determinants of Mineral Ingestion

Ecological niches determine the preferences that animals have for food sources, including minerals (Belovsky, 1978, 1986; Krebs and Davies, 1984). Ungulates are herbivores that are known to search for and then ingest sources of minerals, such as the deposits at mineral springs. Mineral licks are attractive to many such animals (e.g., Weir, 1969; Wiles and Weeks, 1986). For example, elk in Yellowstone National Park will migrate to mineral springs to ingest minerals. In a study in which bighorn sheep in Wyoming ingested at such licks, phosphorus was found at the five salt licks visited, whereas sodium and calcium were found at only three of them. In another study it was suggested that calcium phosphate was an important substance ingested at salt licks, in addition to sodium. Rocky Mountain goats are attracted to salt licks, and their predators (mountain lions) often wait there (Hebert and Cowan, 1971).

Consider the optimization of foraging behavior by the moose. The moose is a herbivore that eats a wide range of brushy plants (Belovsky, 1978). It tends to maximize its ingestive patterns in accordance with the time of day and season, rumen capacity, and energy and sodium requirements. This does not mean that all of its behaviors are correlated in a perfect fashion. But the concordance of operations ensures adequate homeostasis in most instances.

21

There appears to be some evidence of foraging behavior for calcium. The koala feeds almost exclusively on Australian eucalyptus, which does not provide enough calcium for bone development in young mammals. So koalas descend from the trees to ingest mineral-rich dirt (Galef, 1996). Tortoises will ingest bone and will "mine" by digging in soils that are rich in calcium (e.g., Marlow and Tollestrup, 1982; Esque and Peters, 1992). That behavior is especially common among desert tortoises, particularly the females (Marlow and Tollestrup, 1982). And, of course, various species are known to ingest local sources of minerals from dirt, such as chimpanzee (Goodall, 1986) and gorilla (Schaller, 1963), or to migrate to salt licks, perhaps in search of sources of calcium.

## Biological Basis of Mineral/Calcium Ingestion

Migration to salt licks during the reproductive period is a well-known phenomenon in a wide variety of species, as reviewed by Denton (1982). That has its laboratory analogue in the changes in ingestive behavior that are observed during reproduction, including calcium ingestion (Richter, 1943; Denton, 1982); see Chapter 3. Let us consider one example here: The common marmoset, *Callithrix jacchus*, is a small New World primate. It feeds largely on insects, gums, and fruit. During their reproductive years, adult female marmosets are constantly gestating or lactating or both. Their high reproductive rate, combined with a minimal postpartum recovery phase, perhaps leaves them vulnerable to calcium deficiency (Power et al., 1999). In a study to test that idea, marmoset males, nonreproductive females, and lactating females were offered calcium lactate in solution, in addition to their water and food. Lactating females showed greater preference for the calcium solution than did males or virgin females (Figure 1.4). How specific the intake of calcium is during pregnancy and lactation will be addressed in Chapter 3.

## The Nutritionist Tradition

Over the past 50 years, nutritionists, when discussing mineral nutrients, have generally concerned themselves not with the wisdom of the body or the behaviors that support mineral ingestion, but rather with analyzing diets to learn how, when, and why minerals are absorbed by the body once they are consumed (e.g., Institute of Medicine, 1997). Analyses of the compositions of diets and absorptions of required substances have dominated this field (e.g., Oftedal, 1984, 1991). Dietary patterns, and their accompanying absorptions of minerals, vary widely among peoples and cultures (McCance

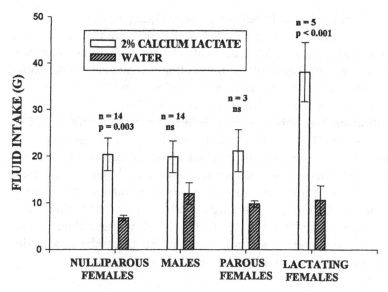

**Figure 1.4.** Ingestion of calcium solution by marmosets over 24 hours. (Adapted from Power et al., 1999.)

and Widdowson, 1942; Prentice, 1994a,b; Kalkwarf et al., 1996; Ritchie et al., 1998). Despite that emphasis, differences in intake, particularly during pregnancy and lactation, have been noted (Prentice, 1994a,b; Ritchie et al., 1998).

During World War II, the importance of maintaining a healthy diet received considerable attention because of food shortages and food rationing. Nutritionists led the way in research to determine what constituted optimum absorption and utilization of minerals such as calcium. The nutritionists' approach to determining the ideal diet included adding substances to the staples of the diet to enhance absorption. That approach – tinkering with the diet to produce optimum performance – became a central tenet of nutrition research, and such recommendations have endured to the present day. The emphasis has been on getting people to "eat right," despite their behavior.

As evidenced by a paper titled "The Inorganic Elements in Nutrition," it has been known for some time that growing and bone-producing animals and milk-producing animals require calcium (Osborne and Mendel, 1918b). An early study of calcium absorption compared the effects of white bread and whole-wheat bread (McCance and Widdowson, 1942). Although the whole-wheat bread contained substantially more calcium, magnesium, potassium, and phosphorus than did white bread, the body was able to absorb calcium from white bread much more readily than from whole-wheat bread. Those researchers concluded that in determining the calcium requirements for a given population, the cereal staple eaten by that population should be taken into account. That study also found that adding vitamin D did not improve calcium absorption, but adding calcium salts did. Therefore, they recommended adding calcium salts to flour in various proportions, depending on whether it was white or whole-wheat flour.

The nutritionists' recommendation was well warranted with regard to dietary self-selection. The problem with the modern nutritionist approach, however, is that behavioral regulation is neglected, perhaps because of the earlier abuse (overreliance on the nutritional-wisdom hypothesis). Moreover, when it comes to humans, culture matters more than anything when it comes to food consumption (Remington, 1936; Mead, 1943; Rozin and Schulkin, 1990; Prentice, 1994a,b; Galef, 1996). Still, vestiges of our evolutionary past and our present mineral needs reveal themselves. Behavioral/biological categories figure importantly.

## Some Evolutionary Thoughts on Calcium Ingestion

For early life forms in the ocean, the emphasis was not on calcium conservation, because the ocean was rich in calcium. Even freshwater typically is rich in calcium (Dacke, 1979). But the ocean was and is an environment in which calcium must be utilized for bodily maintenance (bones) and excreted via gills and renal mechanisms (Wheatley and Greenaway, 1996; Flik et al., 1995). It seems unlikely that a hunger for calcium would have evolved prior to the emergence of land-dwelling animals.

Amphibians, spending part of their time on land, began to evolve storage mechanisms for calcium. One example is the tadpole, which utilizes calcium from storage sacs during bone formation (Dacke, 1979). The calcium sacs are eliminated once the skeleton is formed.

The rise of reptiles and the ongoing colonization of land continued the evolutionary development of physiological regulation of the increasing amounts of bone mass. For most reptiles and birds, calcified eggshell protects against desiccation (Dacke, 1979). Thus there arose increasing needs to acquire, store, and utilize calcium, and calcium appetite may be linked to eggshell production (Dacke, 1979); see Chapter 3. Moreover, the shells are themselves rich sources of calcium, and they often are ingested in a variety of situations.

The transfer of calcium to the fetus in land-dwelling mammals has been an evolutionary factor in calcium regulation (Pitkin, 1975; Packard and Clark, 1996); see Chapter 3. And the need to provide calcium to neonates is imperative for a wide variety of mammals during lactation (Oftedal, 1984). Milk is the primary source of calcium for a neonate, and the mother's bone mass

**Table 1.1.** *Calcium and Phosphorus in Grains*

| Source | Calcium (mg/100 g) | Phosphorus (mg/100 g) |
|---|---|---|
| Uncultivated grains | 132.6 ($n = 119$) | 90.4 ($n = 36$) |
| Cereal grains | 29.1 ($n = 8$) | 288.0 ($n = 8$) |

is an important reservoir from which calcium can be mobilized during that time of increased demand for calcium (Kovacs and Kronenberg, 1997).

Early mammals were predominantly insectivores. According to one hypothesis, the insects that were ingested provided a diet that was higher in calcium than is the diet of modern wild mammals (Eaton and Konner, 1985; Eaton and Nelson, 1991; Heaney, 1993). Thus, during that early period, which may have lasted for 150 million years, our ancestors may have eaten a diet high in mineral content, including calcium.

With the arrival of agriculture, the grains that began to be produced contained less calcium than did uncultivated grains (e.g., wheat, Table 1.1) (Eaton and Nelson, 1991). That situation was reversed with regard to phosphorus.

Consideration of what an animal eats, its behavioral strategies, and its evolutionary history will provide a good indication whether or not that animal will show a calcium appetite. Perhaps many herbivores rarely faced deficits in calcium in their diets, for leafy plants contain large amounts of calcium (Liu and Zhu, 1998).[4] On the other hand, certain species that ingested exclusively fruits and seeds that were not high in calcium may have been vulnerable to calcium deficits (e.g., Dacke, 1979). Moreover, today we see that insect-eating species that ingest the exoskeleton, but not the endoskeleton (where stores of calcium are located), may be vulnerable to calcium deficits. In fact, limestone, which is rich in calcium, is given to gecko lizards in zoos (Dacke, 1979; Allen et al., 1993). Carnivores should not face deficits in calcium if their diets include bones.

## Summary

Evolution has selected behavioral mechanisms in addition to physiological mechanisms to defend the internal milieu. The principles of behavioral regulation (innate and learned behaviors, hedonic responses, salience, etc.) figure into an animal's choices about food and calcium ingestion. There does appear to be an appetite for calcium under conditions of calcium depletion or loss (Chapter 2) and during reproductive periods in females (Chapter 3).

An animal's foraging patterns reflect its biology (Rozin, 1976a), and its foraging patterns are optimizing sets of behavioral and physiological responses (Hainsworth and Wolf, 1990). Palatability factors (Rozin and Fallon, 1987; Berridge, 1996), nutrition considerations (Prentice, 1994a), social learning (Galef, 1986, 1989), and culture (Mead, 1943; Prentice et al., 1983) determine many of our ingestive patterns. Despite such cultural dominance, biological determinants can trigger both physiological and behavioral responses in defense of the internal milieu.

CHAPTER TWO

# Behavioral and Gustatory Regulation of Calcium

This chapter continues with the themes of Chapter 1, but with more detail about calcium appetite and the behavioral and gustatory mechanisms that may underlie it. Calcium deficiency and ingestion of sodium are discussed, in addition to pica and its putative link to calcium deficiency or other deficiencies. Because phosphate is tied to calcium metabolism, there is discussion of the behavioral mechanisms underlying phosphate appetite, although much less is known about phosphate appetite.

There is evidence from a number of different orders (e.g., reptiles, birds, mammals) of increased calcium ingestion when calcium is needed. Nature appears to have selected in favor of behavioral as well as physiological mechanisms for calcium homeostasis. Given this evidence, the question arises whether calcium-seeking behavior is innate or learned, or perhaps some combination of the two. Complicating the question is the fact that whereas sodium has a distinctive taste that allows animals to identify it in foods, that appears to be much less the case for calcium and other minerals.

We begin this chapter with a description of the gustatory system.

## Gustatory Anatomy and Ingestive Behavior

The gustatory system is an essential mechanism in our sensory exploration of the world around us. It is fundamentally linked to ingestive behavior. Gustatory processing is an active sensory function (Gibson, 1966) in the identification of food sources (Pfaffmann, 1967) and is phylogenetically ancient (e.g., Dethier, 1968), sometimes being manifest as gustatory receptors over the body surface, as in aquatic animals (Herrick, 1905). In other words, the gustatory system is an active computational system that assists an animal in adapting to its surroundings.

Gustatory sensibility, olfaction, and oral trigeminal input are considered the chemical senses. The sense of taste plays a primary role in homeostasis (e.g., Pfaffmann, 1967; Bartoshuk, 1991). The oral cavity can be thought of as the outer end of the alimentary canal. It serves to identify substances that are essential for regulation of the internal milieu. Taste contributes to the critical decision to ingest or reject a food, mineral, or fluid source. Gustatory stimuli also trigger gastrointestinal responses, anticipating and facilitating metabolism.

Food selection and utilization require an arousal or interest in a food, the search for it, its recognition, the decision to ingest the food, and, finally, its digestion and absorption (Rozin and Schulkin, 1990). Behavioral solutions to food-selection problems will vary depending on whether the animal is an omnivore (humans, rats, roaches), herbivore, or carnivore, as well as the breadth of its diet (Rozin, 1976a,b).

In the case of sodium, it seems clear that the gustatory system is "programmed" to recognize salty flavors and to promote their ingestion (e.g., Wolf, 1969a; Denton, 1982; Schulkin, 1991c). In the case of calcium regulation at the behavioral level, there may be weaker linkage between the sensory characteristics of calcium in food sources and the ingestion of those sources of calcium. With calcium, there may be a less close relationship between specific gustatory characteristics and the attraction to those sources of calcium.

The tongue is the first contact point for all oral ingesta (Nowlis, 1977; Bartoshuk, 1991) (Figure 2.1). Taste receptor cells in the taste buds transduce the taste stimuli into neural activity that the nerves send to the brain. This occurs at the membrane level via the genesis of chemical synapses. The effects are rapid and reflect transduction mechanisms for gustatory stimuli mediated by G proteins and ionic channels (Roper, 1992; Stewart et al., 1997).

The taste buds are subdivided into five categories, three on the tongue and one each on the palatal epithelium and laryngeal epithelium (Bartoshuk, 1991; Roper, 1992). The taste buds on the tongue and palate monitor most foods, but the purpose of the taste buds on the epiglottis and laryngeal epithelium is unclear, because food and fluids usually do not contact those areas, although perhaps the protection of airways is one function (Travers and Nicklas, 1990; Smith and Hanamor, 1991).

Traditionally, tastes have been divided into four groups: sour, salty, sweet, and bitter (Pfaffmann, 1967; Bartoshuk, 1991). It has been thought that the gustatory system classifies all tastes into one of these four categories. From these gustatory primitive categories, perhaps more complex tastes, such as that for calcium, are constructed.

A mixture of these four basic tastes can result in a number of gustatory signals (e.g., chalkiness). The sweet tastes of foods rich in carbohydrates are associated with nutritious foods, such as fruit. Bitter tastes are associated with potentially toxic substances; quinine is bitter, for example (Janzen, 1977). Salty tastes are associated with sources of minerals, as in a salt lick.

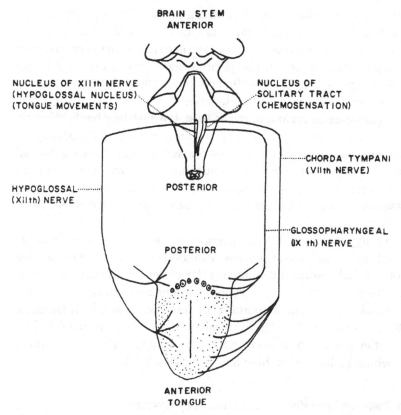

**Figure 2.1.** Anatomy of the tongue. The anterior tongue houses the chorda tympani (the seventh nerve); posterior is the glossopharyngeal (ninth) nerve; also posterior is the vagus (tenth) nerve, also called "the wanderer" because it innervates all the gastrointestinal organs. All three terminate in the nucleus of the solitary tract in the brain to produce chemosensation. (From Nowlis, 1977.)

The function of sour taste may be to stimulate salivary secretion to buffer acids, in addition to serving as a warning signal for spoiled fruit (Rozin and Schulkin, 1990).

But calcium salts produce complex tastes that do not fit neatly into the basic four gustatory categories. This is unlike NaCl, for example, which is the prototypical salt taste (Bartoshuk, 1974; Schulkin, 1982). Instead, calcium salts can elicit a bitter and salty taste (Bartoshuk, 1991; Tordoff, 1996a), a chalky taste (M.L. Power and J. Schulkin, unpublished observations, 1998), or a salty taste (Schulkin, 1982; Tordoff, 1996a).

Three major cranial nerves convey gustatory information from the oral cavity to the brain: Regions of the seventh, ninth, and tenth cranial nerves (Pfaffmann, 1967; Norgren, 1995). Gustatory information is transmitted mostly, though not exclusively, to the anterior region of the solitary nucleus (NTS) (Herrick, 1905; Norgren, 1995). From the nucleus of the solitary tract (in some but not all species that have been studied) there is a major gustatory relay to the parabrachial region. The parabrachial region is subdivided into many smaller regions (e.g., Chamberlin and Saper, 1994), but it appears that the medial region underlies gustatory functions.

From the parabrachial region, in rats at least, the gustatory tracts bifurcate and convey information to the gustatory thalamus and then the gustatory region of the neocortex (insular cortex). A second route is from the parabrachial region to regions such as the central nucleus of the amygdala, the lateral bed nucleus of the stria terminalis, and the lateral hypothalamic region. These areas are known to influence ingestive behavior (Norgren, 1995) (Figure 2.2).

It has been hypothesized that one primary projection trajectory from the parabrachial region underlies the sensory functions of gustatory processing (e.g., to the medial ventral thalamus, insular cortex), while the other gustatory trajectory underlies the motivational and hedonic gustatory processing (e.g., lateral hypothalamus, central nucleus of the amygdala) (Pfaffmann et al., 1977; Spector, 1995). This ventral pathway from the parabrachial region is better construed as a visceral pathway underlying motivational behavior, which includes searching for minerals or food.

## Calcium Taste and the Peripheral Gustatory System

Calcium salts generate strong gustatory signals. They appear to be mediated in part by depolarization of the receptors and reduction of potassium conductance (Roper, 1992), and by calcium-dependent ionic channels (Taylor and Roper, 1994).

Unlike other salts, $CaCl_2$ is autoinhibitory in its effects on chorda tympani firing patterns (Kloub et al., 1998). Beyond 0.3 m, $CaCl_2$ decreases the firing pattern to $CaCl_2$, in addition to those for other salts that have subsequently been tested, including NaCl and KCl. Those authors suggest that because there is inhibition in these other salts, they have certain taste properties in common (Figure 2.3).

To investigate gustatory responses, we tested human subjects and their taste responses to calcium lactate and calcium chloride at several concentrations and where they appeared to be similar in intensity (L. Meldgaard and J. Schulkin, unpublished observations, 1998). Calcium lactate tasted bitter and chalky at the 60-mM concentration, and $CaCl_2$ tasted salty and bitter at the 25-mM concentration.

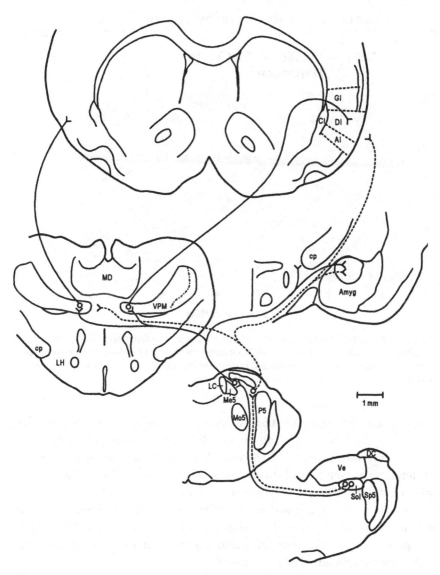

**Figure 2.2.** Schematic summary of the gustatory system in the rat brain. Outlines of coronal sections through the rostral medulla (lower right), pons, diencephalon, hypothalamus and amygdala, and cerebral cortex covering a rostrocaudal distance of about 12 mm. The solid lines connecting the panels represent axons known to convey gustatory information; dash lines represent axons associated with the taste system, but without documented sensory function. None of the lines follow actual pathways, nor do the bifurcations necessarily imply collateral projections. Abbreviations: AI, agranular insular cortex; Amyg, amygdala; CI, claustrum; cp, cerebral peduncle; DC, dorsal cochlear nucleus; DI, dysgranular insular cortex; GI, granular insular cortex; LC, locus ceruleus; LH, lateral hypothalamus; MD, mediodorsal nucleus; Me5, mesencephalic trigeminal nucleus; Mo5, motor trigeminal nucleus; P5, principal sensory trigeminal nucleus; Sol, nucleus of the solitary tract; Sp5, spinal trigeminal nucleus; Ve, vestibular nuclei; VPM, ventral posteromedial nucleus. (From Norgren, 1995, with permission.)

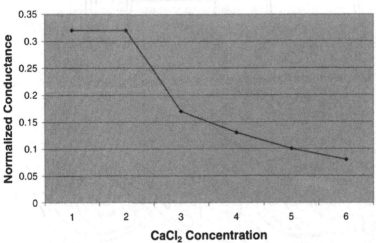

**Figure 2.3.** Normalized conductances at zero current clamp as a function of CaCl₂ concentration. Each point represents a mean ($n = 6$). Note that between 0.1-M and 0.3-M concentrations there is a rapid drop in the transepithelial conductance, whereas a slower drop is observed at a CaCl₂ concentration of 0.3-M. (Adapted from Kloub et al., 1998.)

There is some evidence to suggest that calcium salts can be detected by humans at fairly low concentrations (Tordoff, 1994): Detection of calcium salts was observed between 8 and 50 mM. At those concentrations there were no differences among the solutions that were tested (e.g., CaCl₂, Ca lactate, Ca phosphate). Differences between the two calcium salts were detected at 100 mM, but not at 1 mM. It was noted in that study that as the concentration of calcium salts was increased, so did the unpleasantness. The taste itself is complex and is dependent on concentration. Bitterness, sourness, and saltiness are associated with calcium salts.

In experiments to determine thresholds for calcium salts, rats were trained to detect calcium. The design of the experiment was the following: The rats were trained to discriminate by avoiding shock between water and two different calcium salts (B. Slotnick, L. Meldgaard, and J. Schulkin, unpublished data, 1999). The objective was to determine the lowest concentration at which the rats would give the correct response in at least 75% of 40 consecutive trials. Each of the data points represents the group average over 20 trials; 3 mM was the lowest detectable concentration (Figure 2.4).

Calcium deficiency was found to alter the responses of peripheral (chorda tympani or seventh nerve) gustatory neurons to calcium (Inoue and

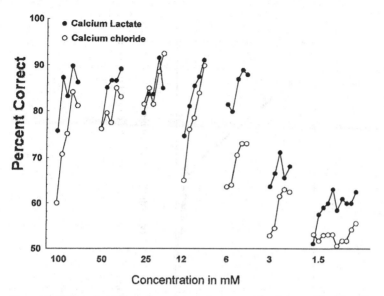

**Figure 2.4.** Threshold data for calcium lactate and calcium chloride. The criterion was set at 40 trials of 75% correct responses or higher. Each animal was run successively on continuously lower millimolar concentration of calcium chloride. As the threshold was reached with one concentration, the next session would start at one concentration lower. Each data point shows the group average on a block of 20 trials. 3.125-mM was the last detectable concentration for the group. (B. Slotnick, L. Meldgaard, and J. Schulkin, unpublished data.)

Tordoff, 1998): Rats were placed on a calcium-deficient diet or on a control diet for 20 days. Calcium deficiency altered the firing rate in the chorda tympani nerve when calcium salts were infused into the oral cavity (Inoue and Tordoff, 1998). When compared with rats on the control diet, rats on the calcium-deficient diet (2 weeks) showed altered firing patterns to low concentrations of calcium chloride and calcium lactate solutions (not shown) and reduced firing patterns to high concentrations, though that effect was quite modest (Figure 2.5). Interestingly, there was no change in firing pattern on infusion of NaCl solution. Those data suggest that changes in gustatory transduction mechanisms at the level of the chorda tympani nerve may influence the calcium ingestion, but not sodium ingestion, that occurs following calcium deprivation. Inoue and Tordoff perhaps rightly suggested that the glossopharyngeal (ninth) cranial nerve might be playing an important role in the transduction mechanisms for calcium taste and calcium appetite. This nerve does not play an essential role in specific sodium ingestion by sodium-depleted rats (Markison et al., 1995).

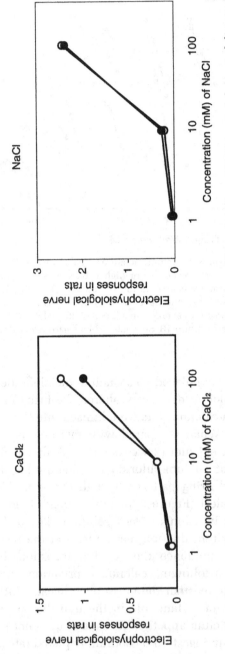

**Figure 2.5.** Chorda tympani nerve responses to 1-, 10-, and 100-mM concentrations of $CaCl_2$ and NaCl in rats fed a control diet or a low-calcium diet. (Adapted from Inoue and Tordoff, 1998.)

## Basic Behavioral Motivation to Ingest Calcium

There are two predominant models that have been used in rats to elicit calcium appetite: parathyroidectomy and feeding a calcium-deficient diet. We consider the parathyroidectomy model first.

Recall that Richter suggested that behavior can serve physiology in regulating the internal milieu. In 1937, Richter and Eckert demonstrated a calcium appetite in rats, and later in rhesus monkeys. In parathyroidectomized animals, calcium retention is compromised, so the rats (the experimental animals in this case) chronically lost calcium. In response, the rats ingested calcium in large amounts, sufficient to fend off tetany. Such rats increased their ingestions of a variety of calcium solutions (acetate, gluconate, and nitrate), as well as strontium and magnesium, which are closely related to calcium chemically. Some of the rats were given parathyroid implants, and their calcium ingestion almost immediately dropped to normal levels.

Some years later the issue was reexamined with regard to the specificity of parathyroidectomy-induced calcium appetite (Leshem et al., 1999a,b). Adult male and female rats were parathyroidectomized. The rats were given free access to food and water. The rats were then offered both calcium chloride and isomolar magnesium chloride. Ingestion was monitored daily. The parathyroidectomized rats increased their calcium intake more than their magnesium intake, and they also reduced their water intake. Another group of rats underwent the same pre-testing procedure, and they were offered both calcium chloride and sodium chloride. There was a preferential increase in calcium intake by the parathyroidectomized rats. Thus, parathyroidectomized rats chose to ingest the calcium salt to a greater extent than either the sodium or magnesium salt.

In another study, it was demonstrated that increased ingestion of calcium in response to parathyroidectomy appeared early in development (Leshem et al., 1999b). In an experiment in which groups of rats were parathyroidectomized at 4, 6, 8, 12 and 20 days of life, all of the rats demonstrated an appetite for calcium (Figure 2.6). In each rat, an intraoral catheter was implanted through which solutions of calcium or magnesium salts could be ingested. Twenty-four hours later, the rats were tested for their calcium ingestion. The rats that had been altered at 4 and 6 days of age had accepted high levels of calcium, but ingested only as much magnesium as did rats that had not been parathyroidectomized. For rats altered at 8 days of age, it was apparent that ingestion of calcium was greater than that of magnesium, and that continued throughout the period in which they were tested. Thus

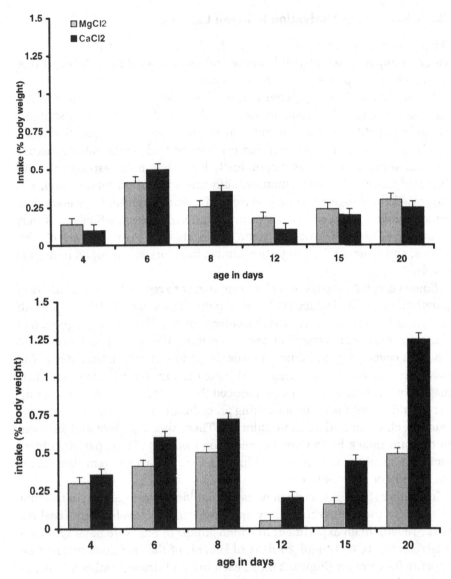

**Figure 2.6.** Intake (percentage body-weight change) of 1.0-M $CaCl_2$ and $MgCl_2$ during 8-minute oral infusions in sucklings, sham (top) and parathyroidectomized (bottom). (Adapted from Leshem et al., 1999b.)

the appetite for calcium is greater than that for magnesium (Leshem et al., 1999b). When newly weaned 26- and 30-day-old parathyroidectomized rats were tested, they also showed preferential intake of calcium salts over magnesium salts (Leshem et al., 1999b).

## Thalamic Gustatory Region and Calcium Appetite Following Parathyroidectomy

One study has suggested that large lesions of the gustatory region of the thalamus will reduce calcium ingestion following parathyroidectomy (Emmers and Nocenti, 1967) (Figure 2.7). In that study, mature rats were parathyroidectomized and were in great need of calcium. Lesions of the thalamic gustatory region were created electrophysiologically by infusing calcium into the oral cavity and recording neuron responses from that region of the brain. Following a recovery period, the lesioned rats showed significantly reduced calcium ingestion as compared with their pre-lesion ingestion of calcium following parathyroidectomy (Figure 2.7).

The lesions most probably interfered with projections from the parabrachial region to the ventral forebrain (central nucleus of the amygdala, bed nucleus of the stria terminalis, lateral hypothalamus), and the behavioral results from the large thalamic lesions may reflect the importance of this pathway (Wolf and Schulkin, 1980; Paulus et al., 1984). This pathway, as noted, is linked to the motivational appetitive component of ingestive behavior (Pfaffmann et al., 1977; Spector, 1995).

## Calcium-Deficient Diet and Calcium Ingestion

Rats placed on a calcium-deficient diet and offered calcium as a fluid or calcium in a food source will opt to ingest the sources of calcium, as demonstrated by a number of investigators (Widmark, 1944; Scott et al., 1950; Tordoff et al., 1990). That phenomenon is quite striking, and the amount of calcium ingested is proportionate to the time the rat has been on the deficient diet (Figure 2.8). In other words, the amount of calcium ingested reflects the duration of the calcium deficiency.

## Appetitive and Consummatory Behaviors

Wallace Craig (1918), an American naturalist, proposed an important distinction between appetitive and consummatory behaviors that is relevant to the study of calcium appetite (Stellar and Stellar, 1985). The appetitive phase has to do with the effort to acquire the calcium. The consummatory phase is the act of ingestion. Thus the question arises whether or not calcium-hungry rats, for example, will seek out calcium (e.g., by pressing a bar to acquire calcium). The answer is that they will. Egg-laying hens (Hughes, 1972) and parathyroidectomized rats (Lewis, 1964) can learn operant responses to obtain calcium. This is a clear example of appetitive behavior, and the degree

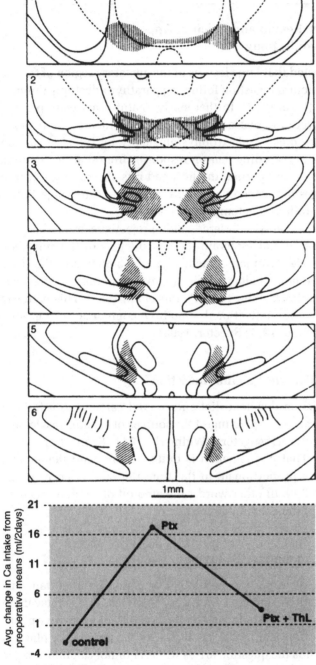

**Figure 2.7.** A representation (top) of a large thalamic lesion within the gustatory region of the thalamus, from anterior (1) to posterior (6). (From Emmers and Nocenti, 1967, with permission.) Calcium ingestion (bottom) by control and parathyroidectomized (Ptx) rats and parathyroidectomized and thalamic-lesioned rats (Ptx + ThL). (Adapted from Emmers and Nocenti, 1967.)

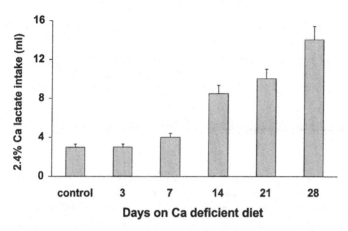

**Figure 2.8.** Short-term (3-hour) intake of 2.4% calcium lactate solution by 5 groups of rats fed a calcium-deficient diet for various durations and by a group of rats fed chow and tested at the same time as the calcium-deprived groups. (Adapted from Tordoff et al., 1990.)

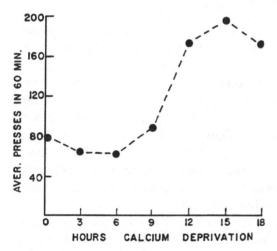

**Figure 2.9.** Extent of bar-pressing for calcium by parathyroidectomized rats deprived of calcium for various durations. (Adapted from Lewis, 1964.)

of calcium deprivation is reflected in the intensity of bar-pressing to obtain calcium, as illustrated in Figure 2.9.

However, in a further study, parathyroidectomized rats were trained in an operant chamber to obtain either calcium or sodium. The rats pressed the bar for the sodium source more frequently than the bar for calcium (Lewis, 1968). In other words, the parathyroidectomized rats pressed both the sodium bar and the calcium bar; it was as if they were not sure about the calcium, so

**Table 2.1.**

| Group | Blood Ca Level (mg/100 ml) | Mean Daily Intake of Ca-Rich Mash as % of Total Food Intake |
|---|---|---|
| Ca-deprived birds | 7.54 (1.3)[a] | 33.06 (2.1) |
| Control birds | 10.6 (1.1) | 22.28 (3.2) |

[a] Standard deviation.

they hedged their bets by ingesting both (Lewis, 1968). Seemingly, either their appetites were not narrowly oriented toward calcium or sodium was also attractive to them. (We will see later that sodium is attractive to calcium-hungry rats.)

## Studies of Calcium Appetite in Chickens

Several studies have shown that calcium-deficient chickens will consume sources of calcium in amounts greater than normal. For example, one study reported that calcium-deprived hens ingested unusual amounts of eggshell (Hellwald, 1931). Another study (Wood-Gush and Kare, 1966) demonstrated that calcium-deprived birds ingested significantly greater amounts of calcium sources in their food than did a control group, and the blood concentrations of calcium were lower in the calcium-deprived group (over a 6-week period) (Table 2.1). In a study of growing chickens given a calcium-deficient diet, the chickens learned to consume a calcium supplement (Joshua and Mueller, 1979). When they were given a diet adequate in calcium, they immediately stopped eating the calcium supplement.

The chicken's appetite for calcium may be a learned behavior resulting from post-ingestion effects (Hughes and Wood-Gush, 1971, 1972). In a series of experiments, chickens were fed a calcium-deficient diet and then given the opportunity to select calcium in different forms (mash and solution) and with variations in appearance (that is, the substances were made to look identical in some cases, distinct in others) (Hughes and Wood-Gush, 1971). For example, red dye was added to calcium carbonate, and calcium-deficient chickens preferred it to an undyed strontium carbonate. When the red dye was put into the strontium carbonate, but not the calcium carbonate, the calcium-deficient chickens ingested the dyed strontium carbonate for some time. After a few days, however, their preference for the red strontium carbonate waned. The researchers concluded that calcium-deprived chickens had a preference for calcium, but that preference was learned as a result of post-ingestion effects,

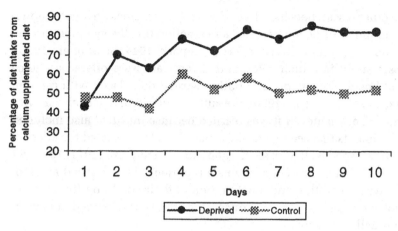

**Figure 2.10.** Daily percentage of calcium intake from a calcium diet by calcium-deprived and non-deprived chickens. (Adapted from Hughes and Wood-Gush, 1972.)

as well as visual and gustatory cues. A further study suggested that the effect that had led to the chickens' learning was an overall, nonspecific feeling of well-being (Hughes and Wood-Gush, 1972) (Figure 2.10).

It is easy to demonstrate avoidance of food sources by animals that would be rendered sick by those foods. It is a ubiquitous phenomenon found in many different kinds of animals (Rozin, 1976a). Moreover, animals that rely on their visual capacities to identify foods also use their visual systems to reject foods (e.g., birds do not have elaborate taste systems, and they use visual recognition to guide their ingestive behaviors) (e.g., Morgan, 1894; Stricker and Sterritt, 1967), and animals that rely on their olfactory senses to select foods use those same senses to reject foods. Learning to avoid a food is easy (Garcia et al., 1974; Rozin, 1976a). Demonstrating that animals can learn to associate the ingestion of specific foods with the amelioration of tissue needs has been much more difficult, as reviewed by Rozin and Schulkin (1990).

## Rat Studies: Is the Appetite for Calcium Innate?

As indicated in Chapter 1, Richter (1943, 1956) postulated that the appetite for calcium, like that for sodium, was innate. Richter speculated that when animals were calcium-hungry, receptors in the oral cavity would be more sensitive to the taste of calcium, and therefore the animals would be more likely to detect and ingest calcium sources. He thought that that occurred through changes in plasma concentrations of calcium impacting gustatory oral sensors.

Several other researchers have hypothesized that there is no innate knowledge of calcium or its beneficial effects in ameliorating the symptoms of deficiency and the presumed discomfort (Widmark, 1944; Scott et al., 1950). In a classic study in which calcium-deficient rats were offered a choice of diets, although the rats preferentially ingested diets laced with calcium (Widmark, 1944; Scott et al., 1950), the authors suggested that such behavior was learned. They believed it was learned because they had also included flavor experiments that seemed to show that the rats associated the effect of calcium with a specific flavor; see Hughes and Wood-Gush (1971, 1972) for similar work in chickens. Other animals reportedly have learned an association between ingestion and amelioration of their condition (Scott et al., 1950). However, the hypotheses concerning innate and learned behaviors are not mutually exclusive.

Experimental findings suggest that early on (by 8 days of age), parathyroidectomized rats will ingest calcium to a greater extent than other solutes (Leshem et al., 1999a,b). Moreover, there appears to be evidence that calcium-deficient rats will ingest a calcium solution within minutes the first time they have access to it (Coldwell and Tordoff, 1996a,b). When offered different solutes in a choice test, the rats immediately showed a positive response to the calcium salts (as well as an increased response to sodium, as explained later in this chapter) (Figure 2.11). That study suggested that the calcium-deprived rats relied on innate oral sensations to identify calcium-containing solutions.

As indicated in Chapter 1, the term "latent learning" describes the acquisition of knowledge under circumstances in which there is no particular motivational state (Tolman, 1949). Latent learning presupposes some innate factors (e.g., recognition of a basic taste quality) (Krieckhaus, 1970). That was particularly true for recalling where sodium was located (Krieckhaus, 1970; Paulus et al., 1984), recalling how to acquire it (Krieckhaus and Wolf, 1968; Dickinson, 1986), and recalling what other tastes might be associated with it (Fudim, 1978; Rescorla, 1981; Berridge and Schulkin, 1989).

In another study, rats were given flavored solutions: calcium solutions paired with one flavor, and sodium solutions paired with another. When the rats were fed a calcium-deficient diet, they increased their intake of both grape- and cherry-flavored solutions, even if those solutions no longer contained calcium or sodium (Figure 2.12). In other words, calcium-deprived rats that earlier had been calcium-replete and were offered solutions of calcium or sodium paired with other tastes would ingest either solution (Coldwell and Tordoff, 1993). By contrast, sodium-hungry rats exposed to either sodium or calcium solutions would ingest only the tastes associated with sodium when they were rendered sodium-hungry for the first time

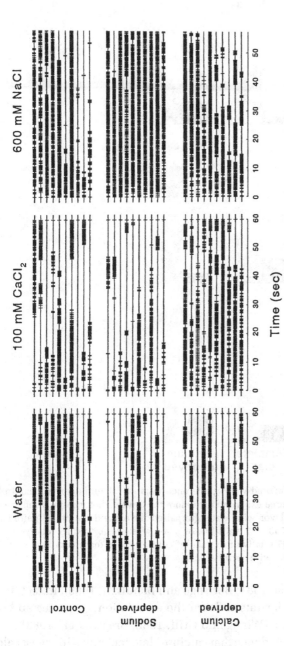

**Figure 2.11.** Licks during the first minute of testing with deionized water, 100-mM CaCl₂, and 600-mM NaCl for groups of Sprague-Dawley rats fed nutritionally complete, low-calcium, and low-sodium diets. (From Coldwell and Tordoff, 1996a, with permission.)

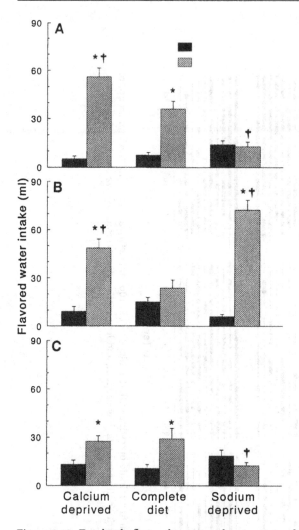

**Figure 2.12.** Two-bottle flavored-water preference tests of calcium-deprived, control, and sodium-deprived rats in three experiments. Before the 3-week diet manipulation, rats were trained to associate one flavor with water (unpaired flavor) and the other (paired flavor) with 100-mM CaCl$_2$ (A), 750-mM NaCl (B), or 584-mM sucrose. *p < .05 relative to intake of unpaired flavor. †p < .05 relative to intake of control group fed a complete diet. (From Coldwell and Tordoff, 1993, with permission.)

(Fudim, 1978; Rescorla, 1981; Berridge and Schulkin, 1989). In fact, it takes just one or two licks for rats to learn the association for places and tastes connected with sodium (Wirsig and Grill, 1982; Bregar et al., 1983).

Thus, there is solid evidence that calcium-deficient rats will ingest calcium salts on first exposure to them, and they will do the same for sodium salts (Coldwell and Tordoff, 1996a,b). In contrast, sodium-deprived rats ingested sodium salts within seconds the first time they were exposed to them, and their appetite was specific (e.g., Wolf, 1969a). The issue of specificity has been a problem for studies of calcium appetite.

The idea of an innate appetite for calcium is quite different from the idea of an innate appetite for sodium. Sodium ingestion is highly constrained to sodium-related ions. Calcium ingestion, at least under some conditions, appears to be much less constrained than the sodium system. Nonetheless, calcium-deficient rats will immediately ingest calcium the first time they are exposed to it; and they demonstrate latent learning. On the latent-learning issue, studies have shown that rats are able to associate two tastes in such a manner that when later one of the tastes is associated with gastrointestinal illness, subsequently both tastes will be avoided (Rescorla, 1981). The issue of latent learning is independent of the issue of innateness, at least to some extent.

## Calcium Deficiency and Ingestion of Sodium

It has been suggested that the evolution of sodium hunger and a specialized system to detect sodium occurred, in part, to regulate other mineral appetites, such as those for potassium and calcium (Schulkin, 1981). Typically, salt licks contain a number of minerals (Hebert and Cowan, 1971; Jones and Hanson, 1985), but the salty taste may be so outstanding that it will attract animals deficient in other minerals. But that hypothesis is limited, for not all sources that contain calcium are salty.

In one study (Schulkin, 1981), 30-day-old rats were placed on a calcium-deficient diet or a thiamine-deficient diet for 21 days. At the end of that period they were given NaCl and various other solutes, including $CaCl_2$. The calcium-deficient rats ingested both the sodium salt and the calcium salts. That was not the case for the thiamine-deficient rats, as shown in Figure 2.13. Interestingly, a study by Okiyama et al. (1996) found that rats fed a low-protein diet would increase their sodium intake but not their calcium intake.

Rats deprived of calcium will ingest sodium (Schulkin, 1981), and the amount of sodium a calcium-deprived rat will ingest will be comparable to or even greater than the amount of sodium an adrenalectomized rat will ingest (Richter, 1936; Tordoff et al., 1990). That demonstrates a robust sodium appetite, for the adrenalectomized rat's sodium ingestion is vital – its life is on the line. Moreover, the amounts of sodium ingested by calcium-deficient rats are pronounced compared with the ingested amounts of other solutions offered, such as citric acid, saccharin, and quinine (Tordoff et al., 1990).

Calcium-deprived rats drank more sodium chloride than did those fed a controlled calcium-replete diet, and they drank more sodium chloride the

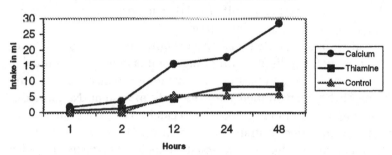

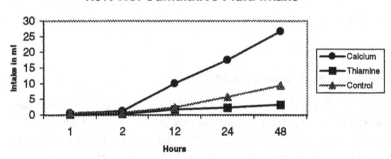

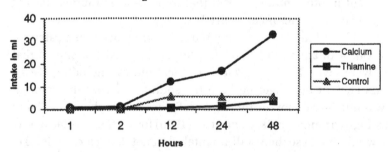

**Figure 2.13.** Ingestions of calcium, potassium, and sodium by calcium- and thiamine-deficient rats and control rats over a 48-hour period. (Adapted from Schulkin, 1981.)

longer they received the calcium-deficient diet (Tordoff et al., 1990). To determine whether the rats responded to only the sodium taste or also to other tastes, two groups of rats, one fed a calcium-deficient diet, and the other a calcium-replete (control) diet, were offered solutions that were sour, bitter, sweet, and salty. Given those options, the calcium-deficient rats drank larger volumes of sodium chloride (salty) than did the control group, beginning at

4–8 days of access, and they increased the amounts they ingested at each opportunity, reaching an average of 127.5 ml/day after 32 days.

Calcium-deficient rats drank somewhat less of the sweet (saccharin) solution than did controls (Tordoff et al., 1990). In general, it has been noted that calcium-deficient rats decrease their ingestion of sweet solutions (McCay and Eaton, 1947; Tordoff and Rabusa, 1998). When saccharin solution was offered, both groups drank very little, although, over time, the calcium-deficient rats drank more saccharin solution than did the controls. Calcium-deficient rats drank slightly more citric acid (sour) than did the controls, although the difference did not become significant until the fourth trial, and intakes did not substantially increase over time. Finally, with a sucrose octaacetate (bitter) solution, the calcium-deficient rats always drank more of the solution than did the control group, but the two groups increased their intake of the solution to the same extent for the duration of the test.

Further studies by Tordoff et al. (1990) evaluated the effects of various calcium supplements on sodium intake by calcium-deprived rats. Groups of rats were fed a calcium-deficient diet and given either no supplement or a calcium carbonate supplement (in varying strengths) or were offered a calcium lactate solution. They had free access to water and sodium chloride solution. Calcium-deprived rats that received no supplements drank significantly more sodium chloride than did any other group. The two groups of rats fed large doses of supplemental calcium (10 g/kg and 50 g/kg) had the lowest intakes of sodium chloride; the intakes for the 10-g and 50-g groups were the same throughout the experiment and did not increase substantially over time.

In that experiment, dietary calcium supplementation affected body weight, food intake, and plasma calcium levels, but not water intake. Rats that received the 10-g supplement gained the most weight, followed by the 5-g supplement group, the 50-g supplement group, the 1-g supplement group, and the unsupplemented group. Plasma calcium levels in the unsupplemented group were below baseline throughout the experiment. Those given 1-g supplements showed low plasma calcium levels at 14 and 21 days, and those given the 50-g supplement had high calcium levels at 14 and 21 days. The 10-g supplement group had high calcium levels at 28 days. When all the rats were given the normal diet, plasma calcium levels returned to normal in less than 48 hours.

The calcium-deficient rats given access to both calcium lactate and sodium chloride increased their intake of calcium lactate for the first 6 days of the experiment, and then that intake plateaued (Tordoff et al., 1990). The effect was similar to that produced by intake of calcium carbonate at about 15 g/kg, and the rats' intakes of food, water, and sodium chloride, as well as weight gain, fell in between the values for the same measures for the rats that received the 10-g and 50-g calcium carbonate supplements. When given a calcium-replete diet, the rats dramatically decreased their intake of calcium lactate.

Subsequent studies by the same group questioned whether or not increased sodium chloride intake in calcium-deprived rats was related to an impaired ability to retain sodium (Tordoff et al., 1993). Those investigators compared the food and water intakes and urine and feces excretions of the calcium-deficient and control rats that were receiving four different-tasting solutions. Compared with controls, calcium-deprived rats drank more water, excreted larger volumes of urine that had lower sodium concentrations, and lost slightly but not significantly less sodium overall. That suggested that calcium deprivation did not disrupt sodium homeostasis and that an appetite for salt did not necessarily reflect a specific need for sodium (Tordoff et al., 1993; Tordoff, 1996b).

As mentioned earlier, studies have shown that parathyroidectomized rats will increase their intake of sodium salts as well as calcium salts. The taste of salt is attractive to calcium-deficient rats. However, in a more recent study, parathyroidectomized rats demonstrated a distinct preference for calcium only, and a decreased intake of sodium (cf. Richter and Eckert, 1938a; Lewis, 1968; Leshem et al., 1999b). Moreover, in many studies, ingestions of calcium by calcium-deprived rats have actually been greater than those of other solutes when they have been offered together, and that includes sodium under most, but not all, conditions (cf. Schulkin, 1981; Tordoff, 1992b; Leshem et al., 1999a,b). It has been suggested (Leshem et al., 1999a,b) that although there is a specific hunger for calcium in the parathyroid-ablated rat, that appetite for calcium may be much less specific during pregnancy and lactation, as will be discussed further in Chapter 3.

## Gustatory Parabrachial Region and Sodium Intake by Calcium-Deprived Rats

For studying the central regions that control sodium intake by calcium-deprived rats, the parabrachial region of the brain stem is a good starting point. The parabrachial region in the brain stem is an important site for a number of ingestive behaviors (Spector, 1995; Norgren, 1995). Electrophysiologically guided lesions of the parabrachial region can interfere with taste-aversion learning (Flynn et al., 1992; Spector, 1995), sodium appetite (Flynn et al., 1992; Scalera et al., 1995), and perhaps calcium appetite (Norgren et al., 1999).

One set of our experiments on calcium appetite was performed in the following way: Gustatory neuron responses within the parabrachial region were electrophysiologically recorded following infusions of tastants into the oral cavity in adult rats. The region was then lesioned, the lesion being largely confined to the medial part of the parabrachial region (Figure 2.14). Following a recovery period, the animals were tested for their responsiveness to solutes during calcium deficiency.

The rats (adults weighing 500–600 g) were placed on a calcium-deficient diet for 3 weeks. Following the 3-week period, they were offered choices of NaCl and $CaCl_2$ in addition to their water and food ad libitum. Concentrated solutions (0.5-M) were used so that the baseline ingestions of the NaCl and $CaCl_2$ would be minimal. The rats were divided into several groups, some of which received the sodium and calcium salts in addition to their normal water and food, and some of which were offered a sodium or calcium solution separately in addition to their water and food. Water ingestion was the same in both groups.

Several findings stand out from that study. First, calcium-deficient rats without the parabrachial lesions ingested the sodium and a little calcium either when given together or when given separately. They ingested more of the sodium solution than of the calcium solution. However, the palatabilities of those two solutes probably were not the same, nor were their concentrations, which might have contributed to the ingestive patterns. Perhaps most important, those were old rats. All of the other studies cited earlier were with developing and younger rats. With this caveat, note that although the lesioned rats did not increase their $CaCl_2$ ingestion, they did increase their NaCl ingestion on the first day of exposure following the calcium-deficient diet (Norgren et al., 1999) (Figure 2.14). In the same study, interestingly, the lesioned rats did not increase their sodium ingestion subsequent to sodium depletion, nor did they demonstrate taste-aversion learning. These results suggest that this region in the brain may be critical for the specific ingestion of sodium when animals are sodium-hungry (e.g., Flynn et al., 1993; Spector, 1995; Norgren, 1995). It is also possible that the fact that the lesioned rats ingested the sodium at all may have reflected a pica response, as discussed later.

## Calcium Deficiency and Reluctance to Ingest Unfamiliar Foods

One behavioral pattern observed even among omnivorous animals is a reluctance to ingest unfamiliar foods. However, nutrient-deficient animals often will ingest any alternative food source (Rozin, 1967b; 1976a). For example, thiamine deficiency in rats, which usually causes anorexia, will induce rats to "immediate, sustained, and vigorous feeding" of novel foods (Rozin, 1976a): Rats were placed on a thiamine-deficient diet for 3 weeks. They were

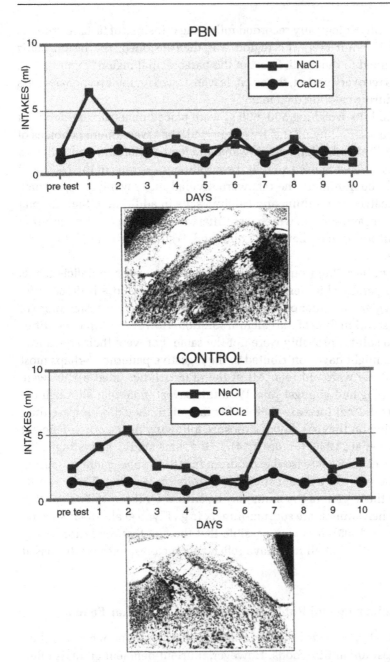

**Figure 2.14.** Sodium and calcium intakes by parabrachial-lesioned rats (PBN) and non-lesioned rats (CONTROL) before (pre) and after being placed on a calcium-deficient diet for 30 days and then offered the salt solutions in addition to water for 10 days. (Adapted from Norgren et al., 1990.)

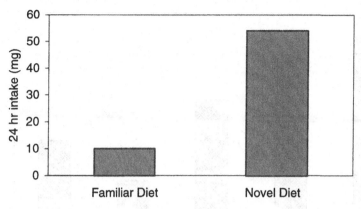

**Figure 2.15.** Ingestion of familiar and novel diets by calcium-deficient rats. (Adapted from Rodgers, 1967.)

then offered a choice between their regular diet and a new, different-tasting diet. For half the rats, thiamine was added back into the familiar diet; for the other half, the familiar diet remained deficient and the new diet was vitamin-enriched. The rats uniformly chose the new diet, regardless of whether or not it contained thiamine. After a few days of eating the new but deficient diet, the rats dramatically reversed course, going back to the familiar diet. Control rats showed no preference for either diet.

Calcium-deficient rats are no different, and like other mineral- and vitamin-deficient rats, they will ingest unfamiliar foods under these conditions (i.e., deficiency). That is, the rats in the laboratory overcome a certain reluctance or natural neophobia during the period of deficiency. When calcium-deficient rats are offered a choice between a diet that is new but not rich in calcium and one that is familiar but only recently enriched with calcium, they will choose the new. In other words, the calcium deficiency results in a preference for a novel diet over a familiar one (Rodgers, 1967). This was demonstrated in an experiment in which rats were fed a calcium-deficient diet for 3 weeks. Those in one group, before the test, were intubated with calcium. The rats were then offered a choice between the same diet and a new diet – sometimes calcium-deficient and sometimes calcium-replete. All groups ingested the novel diet to a greater degree than the familiar diet. In other words, the rats always preferred the novel diet, regardless of whether it contained calcium or not (Rodgers, 1967) (Figure 2.15).

Interestingly, this holds true for vitamin-deficient and other-mineral-deficient rats. However, it does not hold for sodium-deficient rats. They will ingest a novel diet over the old diet, but if the old diet is enriched with sodium, they will return to the old diet (Rodgers, 1967; Rozin, 1976a).

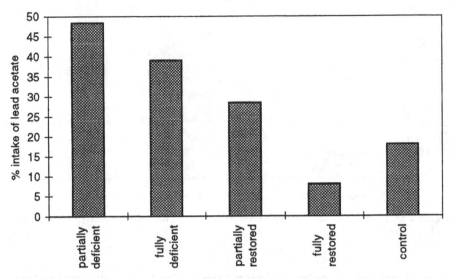

**Figure 2.16.** Lead ingestion as percentage of total fluid intake by rats partially calcium-deficient, fully calcium-deficient, partially restored, and fully restored, as well as control rats. (Adapted from Snowdon, 1977.)

## Pica

Pica is, in part, the ingestion of inappropriate substances, such as laundry detergent, plaster, chalk, paint, or dirt. Geophagia, variously defined as the eating of clay or dirt, is one example of pica (see also Chapter 3). Clay and dirt do contain numerous minerals, and primates and a number of other animals are known to ingest these substances (Schaller, 1963; Weir, 1969; Goodall, 1986).

Several studies have suggested that pica is related to calcium deficiency (Wood-Gush and Kare, 1966; Jacobson and Snowdon, 1976; Snowdon, 1977). There is some evidence that calcium deficiency can elicit pica in both rats and monkeys (Jacobson and Snowdon, 1976; Snowdon, 1977) (Figure 2.16).

A series of experiments evaluated the relationship between calcium deficiency in rats and lead ingestion (Snowdon and Sanderson, 1973; Snowdon, 1977). In one study, rats placed on a calcium-deficient diet were offered a choice between lead acetate solution and distilled water. Whereas control rats appeared to dislike the lead solution, calcium-deficient rats ingested large amounts of the lead. Iron-deficient rats did not (Snowdon and Sanderson, 1973) (Table 2.2). In another study, however, zinc-deficient and magnesium-deficient rats also ingested more lead than did controls, but did not ingest as much as the calcium-deficient rats (Snowdon, 1977).

**Table 2.2.** *Test-Solution Ingestion as Percentage of Total*

|  |  | Lead Solution | | | |
| --- | --- | --- | --- | --- | --- |
| Group | Water | 0.08% | 0.16% | 0.32% | 0.64% |
| Controls (n = 36) | 53.9 | 26.1 | 22.3 | 12.9 | 11.2 |
| Calcium-deficient (n = 27) | 52.1 | 57.4 | 44.6 | 33.0 | 35.6 |
| Iron-deficient (n = 16) | 60.3 | 21.2 | 19.4 | 10.1 | 12.9 |

When calcium-deficient rats were divided into two groups – one given lead acetate solution and one given water – the lead-drinking group gained weight more rapidly than did the water-drinking group, leading those researchers to conclude that lead ingestion may in fact have some ameliorative effects on calcium-deficient rats. When the rats' diets were changed to include adequate amounts of calcium, the rats ceased ingesting high amounts of lead. That study concluded that calcium deficiency led specifically to lead ingestion, but not to other pica behavior. Those researchers reported similar findings in a study of rhesus monkeys (Jacobson and Snowden, 1976). They speculated that lead ingestion in response to the nutritional need for calcium may have mimicked some effect of calcium.

To further study the relationship of lead and calcium, Alheid et al. (1993) gave lead chloride to adult female rats, which were injected with different amounts of lead subcutaneously. They were offered, on alternate days, calcium lactate or sodium chloride, in addition to their water and normal chow. The rats' sodium intake was unchanged, but their calcium intake increased in proportion to the amount of lead they received, findings consistent with the suggested link between lead and calcium ingestion.

The high childhood demands for calcium have led some to hypothesize that childhood pica is linked to calcium needs (Snowdon and Sanderson, 1973; Snowdon, 1977). Childhood encounters with lead are paradigmatic of what we mean by pica – the ingestion of inappropriate things. But in people there have been more examples of pica linked to iron deficiency than to calcium deficiency.

Case studies of three Arab children who engaged in geophagia (they ate sand, stones, and soil) reported that all three had celiac disease (Korman, 1990). Korman suggested that "their pica was a consequence of iron deficiency secondary to celiac disease." A study of geophagia in Tunisian children also found anemia (Karoui and Karoui, 1993). When treated with iron supplements, the children ceased the pica behavior. Those authors compared the pica children with anemic children who did not engage in pica and concluded that geophagia could be simultaneously "the cause and the consequence" of iron and zinc deficiencies. Iron and zinc deficiencies were also the complications in a reported case of coffee-bean pica (Akar et al., 1997), and iron-deficiency anemia was seen in cases of cardboard pica (Callinan and O'Hare, 1988).

It is interesting to note that among children with iron-deficiency anemia, blood levels of selenium (a trace mineral) may also be low (Yetgin et al., 1992). Selenium deficiency has been linked to several diseases (Yetgin et al., 1992) and may be associated with increased risk for cardiovascular disease and some cancers. In a study of anemic children on normal diets and anemic children on generally deficient diets who also practiced geophagia (soileating), the latter had higher levels of selenium. There is no suggestion in the literature that iron deficiency might trigger geophagia, but it is possible that geophagia might have a protective effect against some diseases (Yetgin et al., 1992).

Pica behavior does not always involve toxic or harmful substances. Pagophagia, or ice-eating, is an example. A study of iron-deficient rats demonstrated that, given a choice, they consumed more of their daily water from ice than did control rats (Woods and Weisinger, 1970). When the rats were given sufficient iron, the pagophagia disappeared. Iron-deficiency-related pagophagia is also seen in humans (Parry-Jones, 1992).

## Phosphate Appetite

Phosphate appetite, like calcium appetite, is often linked to bone ingestion (Denton, 1982). It has long been known that cattle eat bone, and it was suggested that such "deranged appetites" of cattle were due to phosphorus deficiency. In the eighteenth century, travelers to South Africa reported that peculiar behavior in cattle, as reviewed by Denton (1982). Such records describe cattle as eagerly searching for and gnawing bones after feeding on what the travelers called "harsh grass" (LeVaillant, 1796). In a similar observation in 1925, Henry Green reported what he called a "perverted appetite" in cattle. Up to 90% of the cattle grazing on certain phosphorus-deficient soils chewed on skeletal bones.

In later studies, investigations revealed that the soil and hence the vegetation in that environment was especially phosphorus-poor, and therefore the herbivores were phosphorus-deprived, which led to a behavioral adaptation to seek bones, the only naturally occurring source of phosphorus

**Figure 2.17.** Phosphate-deficient cow selecting ground-up fresh bone from a cafeteria of trays that also included ground-up old bleached bone, disodium hydrogen phosphate, calcium sulfate (gypsum), calcium carbonate (limestone), and dibasic calcium phosphate. (Courtesy of D. Denton.)

(Denton, 1982). Years later it was demonstrated that phosphorus-deficient cattle would ingest bone when they were offered it (Figure 2.17). The phosphorus deficiency was induced experimentally by a parotid-fistula procedure in which fluids and minerals were drained, and then everything that had been removed, except the phosphorus, was restored to the animal. Further studies indicated that, indeed, an osteophagic phosphorus appetite could be induced in cows by reducing the dietary phosphorus content over months and years (Denton et al., 1986). In another study, ingestion of phosphorus was preferentially chosen from animal bones or sodium-phosphorus solutions (Blair-West et al., 1992).

In another study it took months to induce phosphorus deficiency in a cow, but eventually plasma levels were reduced, and a robust phosphorus appetite was induced. That was in contrast to the findings in several earlier studies that looked at phosphate and calcium deficiencies in cows: There the cows were deprived for several weeks, and the effects on ingestion were negative and inconclusive (Coppock, 1970; Welch et al., 1973; Coppock et al., 1976).

Some long-term phosphate-deficiency studies suggested further that the phosphate appetite seemed specific for bone phosphorus (Blair-West et al., 1992), the ecological niche for the behavioral expression, as bone is the most common source of phosphorus a cow is likely to encounter. Phosphorus deficiency appears to be common in nature, and herbivores may be particularly vulnerable to it. The behavioral adaptation (i.e., eating bone) ensures maintenance of the internal milieu.

A phosphate appetite can be stimulated in juvenile rats, an omnivorous species. Indeed, it was found that the appetite arose rapidly after they were fed a low-phosphorus diet for only 2 days; it appeared to be specific for phosphorus. When offered a choice, the phosphorus-deficient rats ingested more of a phosphorus solution than of a calcium solution (Sweeny et al., 1998). That rapid induction of a behavioral response coincided with the renal adaptation of increased phosphorus reabsorption and may have reflected differential responses from the same general mechanism (Sweeny et al., 1998; Mulroney et al., in press).

## Summary

In summary, gustation characterizes one end of the alimentary canal, providing the important interface between regulation of the internal milieu and one's exploration of the external world to satisfy regulatory requirements. The gustatory behavioral systems underlie calcium appetite. Calcium taste is a complex gustatory signal. Nonetheless, gustatory transduction for calcium has been demonstrated at the level of the peripheral gustatory nerve (seventh nerve). Moreover, though little is known about central gustatory sites, two regions appear to be important: the parabrachial and perhaps efferent gustatory projections to the ventral forebrain.

There seems to be an appetite for calcium in several species in which it has been studied. The appetite appears to be innate, but in certain situations it appears to be less specific than that of the paradigmatic innate sodium appetite (Wolf, 1969a). Calcium-hungry rats can learn operant behaviors to obtain salt and may demonstrate palatability shifts in their perception of calcium salts. Moreover, rats can also remember sources of calcium – where they are and what they are associated with – the first time they are calcium-deficient, despite the fact that the association was made when they were calcium-replete.

Some instances of pica may be examples of calcium hunger (Snowdon, 1977). And having seen that the appetite to ingest calcium in calcium-deficient rats is not foolproof, should anyone be surprised? I think not. In this regard, it is analogous to findings in the literature on human decision-making

(Simon, 1983; Baron, 1994). What one encounters, instead, are biases to ingest certain kinds of calcium sources that perhaps are well grounded but not foolproof. There is no perfect optimization in calcium ingestion. But there is an orientation that is biologically grounded to ensure that calcium can be ingested when it is needed.

Sodium is attractive to calcium-deficient animals. Calcium-deficient rats readily ingest sodium salts, as do rats with several other mineral deficiencies. One strategy for the mineral-hungry animal may be to ingest a wide range of substances, including salty ones (Schulkin, 1981). No doubt more than one strategy is operative in the regulation of calcium metabolism. Perhaps one is to search for something that will satisfy gustatory signal-transduction mechanisms for calcium. The strategy perhaps reflects a specific relationship to calcium ions. In addition, a second strategy may reflect a more broadly based behavioral adaptation in which an animal groups a number of related substances together (salt licks and minerals). It is a well-known phenomenon for a variety of species (Schaller, 1963; Denton, 1982; Goodall, 1986) to be attracted to salt licks, where they can satisfy calcium, phosphorus, or sodium needs.

## CHAPTER THREE

# Gender Differences in Calcium Regulation

Evolution has conferred special behavioral adaptations that are important in the regulation of mineral balance, and the brain generates the behavior of calcium ingestion. This is perhaps most clearly seen in females during pregnancy and lactation.

This chapter will describe some of the effects of the gonadal steroid hormones on ingestive behavior. The effects of estrogen on brain organization during periods of development have long-term consequences for ingestive behavior, including that of calcium ingestion. This chapter will also describe the greater demands for calcium during growth and development, pregnancy, and lactation. Emphasis will be placed on how calcium is regulated during those periods. We will also consider the changes in ingestive behavior that occur in animals and women over the life cycle. In particular, we will examine preferences, aversions, and pica during pregnancy and lactation in women.

## Sexual Dimorphic Effects of Gonadal Steroid Hormones

There are periods during development, both prenatally and postnatally, in which the gonadal steroid hormones have profound effects on body morphology and function. The range of such effects is quite astounding. They include well-known peripheral features (body mass, tissue expression, etc.) in addition to changes in the brain itself (e.g., Gorski et al., 1978).

The structure and neuronal connections of the brain are shaped by activation of gonadal steroid hormones during neonatal development, typically within the first 30 postnatal days (Goy and McEwen, 1977). Regions of the brain such as the medial preoptic nucleus, the medial nucleus of the amygdala, and the medial region of the bed nucleus of the stria terminalis are altered by activation of gonadal steroid hormones during the critical neonatal period (e.g., Simmerly, 1991, 1995). If the male brain is not defeminized by secretion of testosterone and its conversion to estradiol (aromatization process) (McEwen et al., 1977; Goy and McEwen, 1977), it will, in part, develop as if it were female. The production of gonadal steroids during critical periods of development plays a profound role in the organization of the brain and other end organs in the body and in behavior (Goy and McEwen, 1977).

Importantly, however, we now know that gonadal steroid hormones can facilitate structural changes in the brain outside of the narrow window of neonatal development (Arnold and Breedlove, 1985). This is expressed at the levels of the spinal cord and brain stem, and even forebrain regions (Breedlove, 1992; Cooke et al., 1999). But the point to be emphasized is that the steroid hormones have long-term effects on the brain, and they also have periodic effects on the brain, that is, both organizational and activational effects (Arnold and Breedlove, 1985). One way in which steroids alter behavioral events is by inducing peptide or neuropeptide gene expression in the brain (e.g., oxytocin, angiotensin, prolactin) (Herbert, 1993; Pfaff et al., 1997; Schulkin, 1999).

Gender differences in ingestions of fluids and food sources have been well documented. For example, female rats will consume more non-nutritive saccharin or sweet solutions than will males (Zucker, 1969). That behavioral difference between genders begins to appear after puberty (Wade and Zucker, 1969). In addition, researchers have looked at the effects of neonatal hormones, food deprivation, and prior experience on saccharin preference (Wade and Zucker, 1969). In one experiment, female neonatal rats were given either testosterone or estradiol on their fifth day of life. Both hormone treatments resulted in abolition of those rats' sexual cycles and sexual receptivity when they became adults. Compared with control groups, those given testosterone ingested less saccharin; those given estradiol neonatally ingested more saccharin. Comparing the results of those hormonal manipulations with the findings from food-deprivation tests (in which saccharin intake generally increased) and experience tests (in which exposure to saccharin before or after ovariectomy influenced the preference), those researchers concluded that hormonal control was essential for acquisition of the preference, but not for maintaining the preference.

Thus, sex differences in ingestion are influenced by the gonadal steroids *during development*; if females are deprived of estrogen during development, their ingestion of saccharin as adults will be more like that of males; conversely, if males are deprived of testosterone during development, their ingestion of saccharin as adults will be more like that of females.

## Gonadal Steroid Effects on Sodium and Calcium Ingestion

One of the most impressive examples of the sexually dimorphic effects of the gonadal steroid hormones on ingestive behavior concerns sodium ingestion.

Sodium ingestion normally is greater in females than in males in several species in which it has been investigated.

As with saccharin, the gonadal steroid hormones profoundly affect sodium ingestion (Krecek et al., 1972a,b). Removal of estrogen during development (before 30 days of age) renders the female rat more like the male in terms of sodium intake. Conversely, sodium intake by males deprived of testosterone during development becomes more like that for females, that is, elevated salt intake as compared with normal males (Krecek, 1978; Chow et al., 1992; cf. Scheidler et al., 1994).

In fact, female rats show greater positive taste reactions to sodium salts than do male rats. The female's greater avidity for sodium manifests itself in appetitive behaviors, as well as in the consummatory behaviors discussed earlier. One study of male and female sodium-replete rats showed that the females ran faster than the males to reach sodium solutions (P. Arnell, E. Stellar, and J. Schulkin, unpublished observations, 1989). In a study of oral infusions of sodium that looked at characteristic facial patterns accompanying ingestion and rejection, females also showed greater positive oral and facial responses to infusions of salt solutions (Flynn et al., 1993). In such tests, sucrose, for example, typically elicits further ingestion, and quinine leads to rejection. It is clear when a substance is aversive: The rat not only rejects the substance but also demonstrates a characteristic pattern of aversion (Grill and Norgren, 1978; Berridge et al., 1984). There are also indentifiable characteristic facial patterns for positive responses. Thus, at least in rats, females have an innate hedonic bias toward salty tastes (Krecek et al., 1972b; Flynn et al., 1993).

The same pattern of sexual dimorphism holds true for calcium ingestion: Female rats consistently ingest more calcium than do males (Schulkin, 1991a; Leshem and Schulkin, 1998). When adult male and female rats that had adequate levels of calcium in their diets were offered a calcium lactate solution, as well as water (Schulkin, 1991a), within days the virgin females increased their calcium consumption, but male consumption vacillated (Figure 3.1). After both the male and female rats received gonadectomies, the female rats still consumed more than the males (Schulkin, 1991a).

In another experiment, rats were fed a calcium-deficient diet for 10 days and then were offered calcium solutions. Both males and females increased their calcium intake. But the difference in calcium consumption between female and male rats when they were calcium-replete disappeared when both sexes were calcium-deficient (Schulkin, 1991a). When they were once again calcium-replete and they were offered the calcium solution, their ingestion returned to what it had been before the calcium deprivation, and the young females (less than 60 days of age) ingested more calcium than did the males.

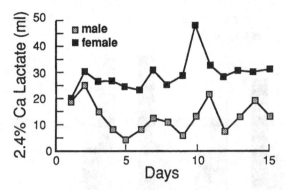

**Figure 3.1.** Daily calcium intake by male and female rats. (From Schulkin, 1991a, with permission.)

As with sodium and saccharin, the sex differences in calcium ingestion during need-free conditions are influenced by the gonadal steroid hormones during development (Reilly and Schulkin, 1993). In another study, neonatal male and female rats were gonadectomized at 1 day of age and thus were deprived of sex hormones during development. At 60 days of age, the rats were given calcium lactate solutions. The gonadectomized males ingested substantially more calcium than did intact male controls. Ovariectomized females ingested substantially less than did intact females. The gonadectomized males and ovariectomized females ingested about the same amounts of calcium solution. We conducted that experiment with several concentrations of calcium and generally found the same trend (Reilly and Schulkin, 1993). The concentrations ranged from 1.2% to 3.6% (Figure 3.2).

## Growth and Development and Calcium Appetite

Among mammals, during early development there is great demand for calcium for bone formation that must be met by the mother's milk. The calcium concentration of mammalian milk is linked to its casein concentration (Oftedal, 1984). The hormones of calcium metabolism are active in the neonate in regulating calcium homeostasis (Pitkin, 1983; Kovacs and Kronenberg, 1997).

When weanling rats were fed a diet very low in calcium, they had a lower dry bone weight than did those fed a high-calcium diet (Fairweather-Tait et al., 1993). That study underlined the importance of adequate dietary calcium during times of rapid growth, because the rats did not appear to have any internal mechanism to compensate for the calcium-deficient diet. Later in life, the two groups of rats had similar calcium absorption rates, suggesting

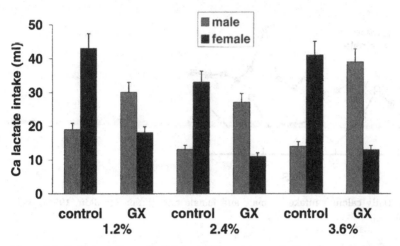

**Figure 3.2.** Daily ingestions of calcium at different concentrations by control male and female rats and rats treated with testosterone at 1 day of age and gonadectomized (GX). (Adapted from Reilly and Schulkin, 1993.)

that calcium deprivation in early life does not affect an animal's ability to absorb calcium in adulthood.

Neonatal female and male rats will increase their calcium ingestion when calcium-hungry. As discussed in Chapter 2, neonatal rats as young as 6 days of age were seen to increase their calcium intake when it was infused into the oral cavity after parathyroidectomy; the ingestion appeared to be preferential for calcium salts by 8 days of age (Leshem et al., 1999a). Importantly, there were no differences in calcium consumption between male and female neonates when they were rendered calcium-hungry by parathyroidectomy, nor between male and female young adults when they were calcium-deprived (Schulkin, 1991a). Whereas a neonate will normally receive its calcium through the mother's milk, which is rich in calcium, these experiments suggest that the brain is prepared for selective calcium ingestion or perhaps mineral ingestion. More generally, it is known that neonates can increase their intake of water, sodium, and food when tested during the first week of life (e.g., Hall, 1979; Leshem et al., 1988).

## Extracellular-Fluid Regulation of Water and Sodium and Its Relationship to Estrogen Levels

It appears that female rats are more sensitive to alterations in extracellular-fluid balance than are males. Both water and sodium ingestions are essential to maintain the body's fluid volume, and female rats are more likely than males to ingest more water, as well as more sodium, when they are hypovolemic (Fitzsimons, 1979, 1998; Kaufman, 1980).

It has been found that levels of estrogen will influence the dipsogenic action of angiotensin administered to elicit water intake (Vijande et al., 1978; Kisley et al., 1999) and sodium ingestion (Fregly and Thrasher, 1978; Danielson and Buggy, 1980). Female rats reduced their water-intake responses to angiotensin during estrus, when estrogen levels peaked (and they consumed less saline solution), whereas when males and females that had been gonadectomized at birth were given angiotensin, they drank about the same amounts of water (Vijande et al., 1978; Findlay et al., 1979; Danielson and Buggy, 1980).

A similar study found that the cyclical changes in water intake seen in the ovulating female in response to angiotensin were not observed with hypertonic-saline-induced thirst (Findlay et al., 1979). That study concluded that the thirst of extracellular origin is dependent on activation of the renin-angiotensin system (Findlay et al., 1979). Another study confirmed that finding: Carbachol-induced thirst did not change when female rats were given estradiol (Jonklaas and Buggy, 1985a,b). Again, water ingestion that does not reflect extracellular-fluid regulation is not altered during the estrous cycle.

Scheidler et al. (1994) showed that rats injected with estrogen would ingest less sodium under a number of conditions, including sodium depletion induced by furosemide or adrenalectomy. Removal of the ovaries in adult rats abolished that effect.

In sheep, sodium intake increases during the luteal phase, but decreases on the days following estrus (Michell, 1975). However, the sheep's peak sodium loss occurs at estrus, when sodium intake is lowest (Michell, 1976). There are also differences in sodium intake that are linked to individual differences in sheep during the estrous cycle. We do not know if intake of calcium would be altered along the same lines as that of sodium and other solutes. We do know that in chickens, neither estrogen infusions nor testosterone infusions reduced calcium ingestion (Lobaugh et al., 1981).

Finally, estrogen levels are known to influence food intake. During estrus, food intake decreases (Tarttelin and Gorski, 1971). There is some evidence that whereas protein intake increases with elevated estrogen levels, fat intake decreases (Bartness and Waldbillig, 1984). These effects are mediated by the preoptic region, and perhaps by induction of oxytocin or cholecystokinin gene expression (e.g., Butera et al., 1996). Estrogen implanted in that region induces the decreases in food, water, and salt intake described earlier. Estrogen implanted more caudally in the ventral medial region does not produce these effects on ingestive behavior (Jonklass and Buggy, 1985a,b).

## Menstrual Cycle, Estrogen Levels, and Ingestion

Changes in ingestive behavior during the menstrual period have been well documented. Generally, there is a decrease in appetite, and ingestions of a number of substances are reported to decrease during menstruation, including salts. However, there seems no clear association of a food source and aversion during this period (e.g., Buffenstein et al., 1977; Tomelleri and Grunewald, 1987; Bancroft et al., 1988).

A study conducted with rhesus monkeys found an increase in food rejection during the ovulation phase of the menstrual cycle (Csaja, 1975). The study also found increased food rejection among the monkeys during the early weeks of pregnancy, similar to the experiences of nausea and appetite changes that accompany the early stages of human pregnancy. That study suggested that estrogen levels might underlie the changes in ingestion. Other studies have suggested that parathyroid hormone might underlie the decreased appetite during the premenstrual period (Thys-Jacobs and Alvir, 1995).

Carbohydrates have notable effects on women during the menstrual cycle (Sayegh et al., 1995). In general, sweet tastes are strongly preferred during menstruation (Bowen and Grunberg, 1990), perhaps because of the immediate energy they provide (sweet-tasting carbohydrates).

Studies evaluating women's food cravings or changes in food preferences over the menstrual cycle have reported increased intake of sweets and/or chocolate, either during menses or during the premenstrual period (the 2 weeks prior to menses) (Tomelleri and Grunewald, 1987; Bowen and Grunberg, 1990; Rozin et al., 1991). One researcher hypothesized that age affects the timing of the craving for sweets: Younger women (21 years and under) craved sweets during menses, and older women (31 years and over) craved sweets during the premenstrual phase (Moos, 1968).

Although there does not appear to be an increase in preference for salty foods in relation to the menstrual cycle (Bowen and Grunberg, 1990; Kanarek et al., 1995), one study has noted changes in the palatability of salty food (specifically popcorn) over the menstrual cycle (Frye amd DeMolar, 1994). That study also found that men had higher preferences overall for the various levels of salt in popcorn than did women (Frye and DeMolar, 1994).

We looked at calcium responsiveness across the menstrual cycle among undergraduate university students and found that neither intensity grading for calcium nor preference for calcium was altered over the menstrual cycle (M. Leshem, T. Levin, and J. Schulkin, unpublished observations, 1997).

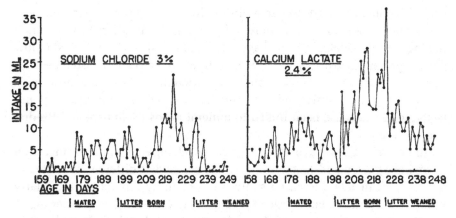

**Figure 3.3.** Curves showing average daily intakes of sodium and calcium before mating, during pregnancy, during lactation, and after weaning. (Adapted from Richter, 1955.)

## Pregnancy and Lactation: Animal Ingestive Studies

Several species are well known to have greater needs for calcium and to be calcium-challenged during pregnancy (e.g., brown bat) (Barclay, 1995; Hood, 1998). Few, however, have been tested for a calcium appetite.

There are well-known changes in the hormonal milieu in most animals that have been studied during pregnancy and lactation. Curt Richter (1943) demonstrated that during pregnancy and lactation and following weaning there were systematic changes in ingestive behaviors (Figure 3.3). For example, he showed that fat intake and protein intake increased, but carbohydrate intake did not. He also showed that in rats, sodium intake increased during pregnancy and lactation and returned to normal following the weaning period. Because the rats increased their calcium and sodium intake within days of mating, Richter speculated that the appetite had maternal and not fetal origins. The sodium effects have been noted by other observers (Pike, 1971), but not by all (Thiels et al., 1990).

Pregnant mice are known to increase their ingestion of sodium and to continue to do so during lactation, and there have been reports of increased salt ingestion by sheep during pregnancy (Denton, 1982), though others have not found such increased intake during that period (Michell and Moss, 1988).

Among female rats, increased genital licking has been reported during pregnancy and lactation; that could be a behavioral mechanism used to recycle sodium (Steinberg and Bindra, 1962; Denton, 1982). Licking of rat pups has been construed as part of the ingestive behavior of the mother, in which water and minerals are essential for metabolization during the reproductive cycle (Friedman et al., 1981). In addition, ingestion of the feces of the newborns may be an important source of calcium, in addition to the genital licking of the neonates. Coprophagy (eating of feces) has been linked to low dietary protein (Rozin, 1976a). It is not known if feces are ingested because they are sources of minerals, though the secretion of calcium in feces is significant. But certainly genital licking has been observed in pregnant and lactating rats and has been linked to mineral needs (Steinberg and Bindra, 1962; Denton, 1982).

Studies of rabbits have shown dramatic changes in mineral ingestion during pregnancy and lactation, although somewhat different appetites in rabbits than in rats. Increased sodium ingestion by rabbits is particularly pronounced (but at a later stage of pregnancy than in rats), as is water ingestion. Calcium ingestion by rabbits is also increased significantly, and within days of mating. It was found that magnesium and potassium intakes did not change (Denton and Nelson, 1971) (Figure 3.4). On the other hand, in rats, we found that magnesium ingestion was increased during pregnancy and lactation, and to a greater extent than that of calcium (Leshem and Schulkin, 1998).

During lactation, rabbits' sodium intake increases dramatically. The greater the litter size, the greater the sodium ingestion. In contrast, calcium ingestion is about the same as that during pregnancy, and litter size does not appear to have an effect (Denton, 1982).

In rabbits, infusions of prolactin resulted in increased calcium ingestion and sodium ingestion, whereas infusions of oxytocin increased sodium ingestion but did not affect intake of calcium or other solutes (Denton, 1982). Interestingly, prolactin has been linked to calcium metabolism and is known to elicit increased food and water ingestion by lactating rats (Kaufman et al., 1981; Kaufman and Mackay, 1983) and chickens (Buntin, 1992). Those findings led researchers to conclude that prolactin has a priming effect on the mechanism of calcium appetite (Figure 3.5).

Interestingly, a prior pregnancy results in greater sodium ingestion during a subsequent pregnancy (Frankmann et al., 1991). A similar phenomenon holds for calcium ingestion (Reilly et al., 1995). Multiparous female rats ingest more calcium than do virgins (Reilly et al., 1995) (Figure 3.6).

Egg-laying chickens also choose to ingest calcium (Hughes, 1972). They actually seem to ingest the calcium to a much greater extent in anticipation of calcium requirements (Hughes, 1972). During ovulation, the average ingestion of calcium was increased over that seen when not ovulating (Figure 3.7) (Hughes, 1972), suggesting that the increase in calcium consumption was in anticipation of calcium requirements during the calcification of the shell.

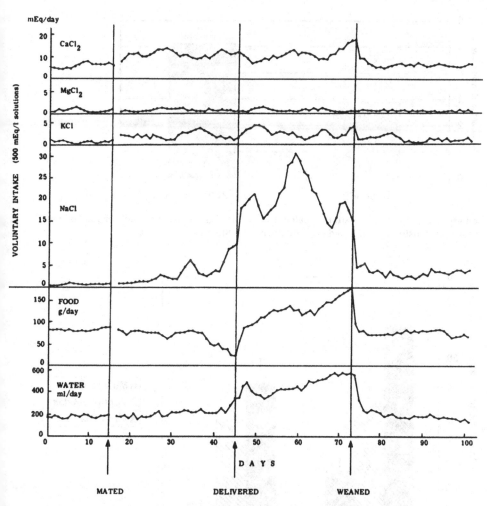

**Figure 3.4.** Mean daily intakes of food, water, and 500-mEq/liter solutions of NaCl, KCl, MgCl$_2$, and CaCl$_2$ by 7 rabbits during a control period, during pregnancy and lactation, and subsequent to lactating. (From Denton and Nelson, 1971, with permission.)

Low-calcium diets result in smaller eggs with thinner shells. When chickens are given a diet with no calcium, egg-laying ceases (Elaroussi et al., 1994). During the egg-production period, hens lose a considerable amount of calcium (Dache, 1979): The calcium in an average-size eggshell is equal to 10% of the hen's total body calcium. The average commercial egg-laying hen deposits 30 times her total body calcium in eggs over one year (Elaroussi et al., 1994).

In studies using rats, vitamin D deprivation resulted in compromised ability to retain, reabsorb, and utilize calcium, and under those conditions

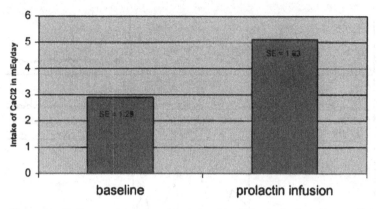

**Figure 3.5.** Daily mean ingestions of calcium by rabbits during baseline conditions and during prolactin infusion (50 I.U., 10 days). (Adapted from Shulkes et al., 1972.)

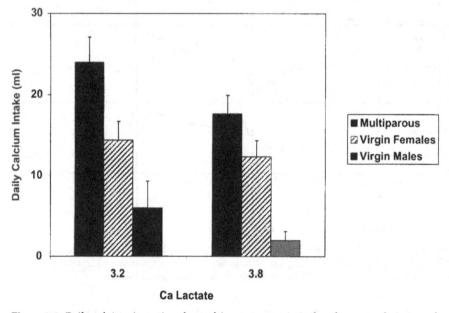

**Figure 3.6.** Daily calcium ingestions by multiparous rats, virgin female rats, and virgin male rats. (Reprinted from *Brain Research Bulletin* 37, J. J. Reilly, J. Nardozzi, and J. Schulkin, "The ingestion of calcium in multiparous and virgin female rats," pp. 301–3, copyright 1995, with permission from Elsevier Science.)

there was greater calcium loss. When deprived of vitamin D during lactation, female rats self-selected a high-calcium diet (from a choice of three diets high in calcium, high in phosphate, or normal in calcium and phosphate) (Brommage and DeLuca, 1984). Those rats avoided the high-phosphate diet,

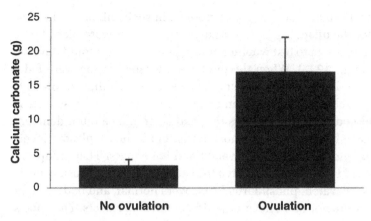

**Figure 3.7.** Daily calcium ingestions by ovulating and non-ovulating chickens. (Adapted from Hughes, 1972.)

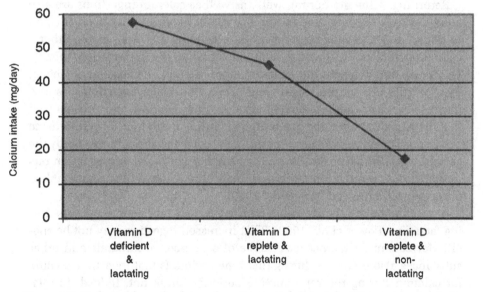

**Figure 3.8.** Daily intakes of calcium by vitamin-D-deficient lactating rats, vitamin-D-sufficient lactating rats, and vitamin-D-sufficient non-lactating rats. (Adapted from Brommage and DeLuca, 1984.)

preferring to ingest the high-calcium diet. The high-calcium diet consumed by the vitamin-D-deficient lactating rats resulted in increases in plasma calcium levels, increased food consumption, and stimulation of milk production. Calcium intake by vitamin-D-deficient lactating rats clearly was greater than that by vitamin-D-sufficient non-lactating rats (Figure 3.8).

The ingestion of calcium by pregnant mothers in such animal studies has implications for the offspring. In one study, pregnant rats were placed on a vitamin-D-deficient diet to test whether or not their offspring would develop rickets (Clark et al., 1987). When the pups were weaned, some were fed a calcium-deficient diet, and others were fed a calcium-rich diet. At 56 days of age, plasma levels of phosphorus and calcium were measured, as well as bone mass and body weight. The rats that had eaten the calcium-deficient diet had normal plasma levels of phosphorus, but reduced plasma levels of calcium and below-normal body weight and bone mass. That group did develop rickets. In the rats that had eaten the calcium-rich diet, plasma levels of calcium were elevated, phosphorus levels were normal, and body weight and bone mass were normal. That group did not develop rickets. The effects of vitamin D deficiency were overcome by the effects of increased calcium. Plasma calcium levels returned to normal in the vitamin-D-deficient parental lactating group after ingesting calcium.

Returning to the mother rat, water as well as calcium ingestions are also altered during reproduction (Kaufman, 1980). In one study (Woodside and Millelire, 1987), groups of pregnant and lactating rats were given either a calcium-sufficient diet or a calcium-deficient diet. During lactation, the rats on the calcium-deficient diet ingested more calcium (2.4% calcium lactate) than did the rats on the calcium-sufficient diet. Water ingestion was also increased in the calcium-deficient lactating rats. A follow-up study showed that the larger the litter, the greater the ingestion of calcium. That difference in calcium intake was not seen in the first week of lactation, but it was by the second week (Millelire and Woodside, 1989) (Figure 3.9). Moreover, in rats in which milk production decreased, ingestion of calcium was nonetheless observed (Millelire and Woodside, 1989).

As mentioned in Chapter 1, common marmosets ingest more calcium during lactation (Power et al., 1999). That increased ingestion may not be specific for calcium salts, because intake of a number of minerals and other substances also increases during that time period. Is that ingestion specific for calcium during the reproductive period? I think not. Indeed, in rats, calcium ingestion increases during pregnancy and lactation, but so does ingestion of other substances, which seems more like evidence of pica than of the body's regulatory wisdom. Even quinine ingestion can be increased during that period; on the topic of substances usually aversive to rats, see Leshem and Schulkin (1998). Thus, the appetites that emerge during pregnancy and lactation in rats may reflect changes in taste acuity, in addition to the desire for specific nutrients and minerals (Lewis, 1968; Leshem and Schulkin, 1998; Levin et al., 1999) (Figure 3.10).

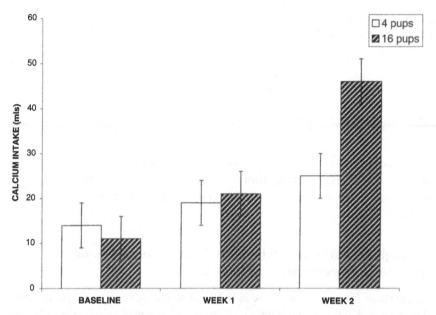

**Figure 3.9.** Daily intakes of 2.4% calcium lactate solution under baseline and lactating conditions by rats with litters of 4 and 16 pups. (Adapted from Millelire and Woodside, 1989.)

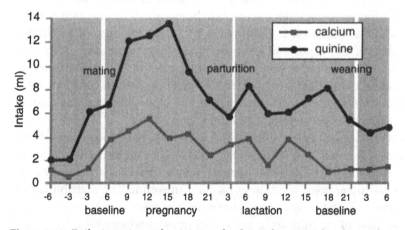

**Figure 3.10.** Daily ingestions of quinine and calcium by rats at baseline, when pregnant, when lactating, and following weaning. (M. Leshem and J. Schulkin, unpublished data.)

In a further study, pregnant multiparous rats demonstrated increased calcium intake, but they did not show greater positive oral and facial responses to the infusions of calcium. On the other hand, their increased consumption of quinine was associated with fewer negative oral and facial responses to infusions of quinine (Levin et al., 1999).

**Table 3.1.** *Distribution of*
*Calcium in a Pregnant Woman*

| | |
|---|---|
| Fetus | 28 g |
| Placenta | 6 g |
| Amniotic fluid | 1 g |
| Extracellular fluid | 3 g |

Thus, there are changes in ingestive behaviors during pregnancy and lactation, but the ingestion of calcium in animal studies that is known to occur is not, in my view, specific for calcium per se. A wide range of changes in ingestive behavior can occur.

## Physiological Interlude: Calcium Metabolism and Hormonal Regulation in Pregnant and Lactating Women

An important topic for this book is calcium metabolism during pregnancy and lactation. Therefore we must consider the changes in calcium-regulating hormones that occur in humans during those periods. The hormones of calcium homeostasis will be discussed in more detail in the next chapter.

Many studies have documented calcium homeostasis in pregnant women (Heaney and Skillman, 1971; Dahlman et al., 1994; Institute of Medicine, 1997), although the reports of changes in the concentrations of specific calcium-related substances vary widely. Total calcium decreases throughout pregnancy, but plasma levels of ionized calcium remain constant. After giving birth, the mother's total calcium begins to increase, reaching normal levels by 2 months after delivery (Dahlman et al., 1994).

In pregnant women, extracellular-fluid volume expands, and renal function increases (Kovacs and Kronenberg, 1997). Calcium from the mother is transported to the fetus. Maternal estrogen levels increase, which may contribute to the lower concentration of calcium in serum. The distribution of calcium stores during pregnancy is shown in Table 3.1.

The transport of calcium from mother to fetus produces mild hypercalcemia in the fetus. At term, the total intrauterine calcium accumulation is 25–30 g, which is the same as in the fetal skeleton. On average, calcium is transferred to the fetus during the last half of pregnancy at a rate of 200 mg/day, when fetal skeletal mineralization takes place (Institute of Medicine, 1997).

Parathyroid hormone (PTH) and calcitonin do not appear to cross the placenta (Kovacs and Kronenberg, 1997). Instead, the placenta has its own sources of PTH, calcitonin, and vitamin D (Kovacs and Kronenberg, 1997).

The placenta transports calcium ions from fetus to mother and from mother to fetus. Perhaps this occurs via calcium-transporting mechanisms induced by vitamin D and PTH, possibly even at the level of the placenta itself (Kovacs and Kronenberg, 1997). The mother is funneling calcium to the fetus, regardless of the fetus's need. Because the fetus is hypercalcemic, it has low levels of PTH (Kovacs and Kronenberg, 1997). This active transport of calcium ions through the placenta is one mechanism for maintaining a positive fetal calcium balance (Pitkin, 1975, 1983).[1]

Calcium concentrations in cord blood are higher than maternal concentrations, again reflecting the higher concentrations of calcium ions in the fetus. PTH or PTH-related peptide is present in the fetus from about 10 weeks of age, and the presumed response to fetal relative hypercalcemia would be decreased PTH production (Kovacs and Kronenberg, 1997).

There is some confusion between the older literature and the new literature concerning precisely when PTH is elevated during pregnancy and in the immediate postpartum period. Only recently has PTH-related peptide been discovered. It is distinct, but shares some properties with PTH, as reviewed by Kovacs and Kronenberg (1997). With this caveat in mind, we know that maternal PTH-related peptide, but not PTH, appears to be increased during pregnancy. That might explain the earlier confusing and conflicting findings about PTH levels (cf. Pitkin, 1975, 1983; Kovacs and Kronenberg, 1997). We can conclude that PTH levels remain low until the midpoint of pregnancy and then increase through the postpartum period. Calcium balance and levels of hormones are depicted in Figure 3.11, which summarizes data from a number of studies.

Recent evidence suggests that calcium excretion is increased during pregnancy, but then decreases during the early part of lactation (Klein et al., 1995; Seely et al., 1997; Ritchie et al., 1998; cf. Pitkin, 1983). The differences in findings may reflect individual variations in calcium balance. However, plasma calcium concentrations remain almost constant, and calcium reabsorption from the intestine is increased.

No doubt the mother's greater demand for calcium during pregnancy reflects the fetus's greater demand for calcium. As shown in Figure 3.11, concentrations of vitamin D ($1\alpha,25$-dihydroxyvitamin $D_3$), PTH-related peptide, and ca citonin are increased during pregnancy (Seely et al., 1997; Ritchie t al., 1998).

A study that looked specifically at intestinal calcium absorption during lactation and weaning hypothesized that calcium absorption and calcitriol vitamin D metabolite) serum levels are higher during those periods to compensate for calcium lost during breast-feeding (Kalkwarf et al., 1996). Those researchers concluded that intestinal calcium absorption did not become

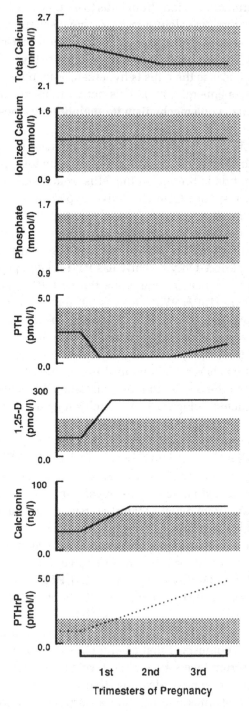

**Figure 3.11.** Schematic illustration of the longitudinal changes in calcium, phosphate, and calcitropic hormone levels that occur during human pregnancy. Normal adult ranges are indicated by the shaded areas. The progression in PTHrP levels is depicted by a dashed line to reflect that the data are less certain. [From C.S. Kovacs and H.M. Kronenberg, "Maternal-fetal calcium and bone metabolism during pregnancy, puerperium, and lactation," *Endocrine Reviews* 18:832–72, 1997, © The Endocrine Society, used with permission.]

more efficient during lactation, but only when menses resumed or after weaning (Kalkwarf et al., 1996). In general, calcium supplementation had no significant effect on calcium absorption during those periods of reproduction. Among women in the lactation group, those who had resumed menses had a fractional calcium absorption measurement of 0.39, compared with 0.30 among those who had not. Lactating women who had weaned their babies had an absorption measurement of 0.37, compared with 0.31 for non-lactating women. There was a net 46-mg/day increase in total dietary calcium absorbed by lactating women who had weaned. Serum levels of 1,25-(OH)$_2$-D$_3$ followed along the same lines as calcium absorption: higher for lactating women who had resumed menses, and also higher for lactating women who had weaned.

Other studies have analyzed daily excretions to determine whether or not postpartum women conserve urinary calcium, magnesium, or zinc (Klein et al., 1995). Lactating women were compared with postpartum non-lactating women and women who had never been pregnant. The subjects' intakes of calcium, magnesium, and zinc were measured and generally were stable. The lactating women excreted less calcium than did the non-lactating and never-pregnant women.

To determine the roles of PTH, calcitonin, and vitamin D in lactation, researchers studied women who were breast-feeding twins, because they secreted more milk than women breast-feeding single infants (Greer et al., 1982c). Serum levels of the various substances were measured by blood tests. Both groups had increased serum concentrations of calcium, but the mothers nursing twins had higher concentrations. Mothers of twins also had significantly higher levels of PTH and calcitonin than did mothers of single-tons. Levels of vitamin D were also higher in the mothers of twins, but the significance of the difference declined over time. Mothers of twins had significantly higher caloric intakes, with significantly higher calcium intakes in particular. Those researchers concluded that mothers of twins compensated for their increased losses of calcium in breast milk by increasing their dietary intake of calcium and by increasing their serum concentrations of PTH, calcitonin, and 1,25-(OH)$_2$-D$_3$.

Other investigators have evaluated PTH-related peptide (PTHrP) in relation to lactation (Sowers et al., 1996). Groups of postpartum women who were fully breast-feeding, partially breast-feeding, or bottle-feeding their infants were monitored at intervals for 18 months. Researchers measured serum levels of prolactin, PTHrP, PTH, estradiol, and vitamin D by blood tests and compared bone-mineral densities at the femoral neck and the lumbar spine using dual-energy x-ray absorptiometry. Women who breast-fed exclusively had significantly higher levels of PTHrP than did those who bottle-fed,

**Table 3.2.** *Factor Changes during Pregnancy and Lactation*

| Factor | Human | Rat |
|---|---|---|
| During pregnancy | | |
| Ionized calcium in blood | Stable | Reduced in late pregnancy |
| PTH | Low to low-normal from early pregnancy | Increased |
| 1,25-(OH)$_2$-D$_3$ | Increased in early pregnancy | Increased in late pregnancy |
| Intestinal calcium absorption | Increased; follows rise in 1,25-(OH)$_2$-D$_3$ | Increased; precedes rise in 1,25-(OH)$_2$-D$_3$ |
| During lactation | | |
| Ionized calcium in blood | Stable or slightly increased | Reduced |
| PTH | Low to low-normal | Increased |
| 1,25-(OH)$_2$-D$_3$ | Normal | Increased |
| Intestinal calcium absorption | Normal | Increased |
| Skeletal calcium losses | 3–8% | 30–35% |

and as breast-feeding was tapered off, so did PTHrP levels. High PTHrP levels were also associated with high prolactin levels. Women who breast-fed exclusively had lower levels of estradiol than did those who bottle-fed.

The onset of lactation is triggered by decreases in estrogen and progesterone levels and by elevation of prolactin. Prolactin and oxytocin levels are elevated for lactation and perhaps also to promote mother-infant attachment (e.g., Bridges et al., 1990). Figure 3.12 depicts changes in calcitropic hormones and calcium and phosphate levels during lactation, during weaning, and post weaning (Kovacs and Kronenberg, 1997).

PTH levels are quite high in human milk, much higher than in plasma (Institute of Medicine, 1997). These levels of PTH are associated with lactation and with lactation-related decreases in bone density of up to 6% (Prentice, 1994a; Kalkwarf et al., 1997). Low estrogen and high PTH concentrations are linked to bone demineralization at that time, and concentrations of PTH have been correlated with density changes in the lumbar spine and the femoral neck (Institute of Medicine, 1997).

Finally, we should keep in mind the differences between rodents and people with regard to the calcitropic hormones (Kovacs and Kronenberg, 1997). Let us next return to the issue of behavior. In the next chapter we will discuss the putative link between the hormones of calcium homeostasis and calcium ingestion (Table 3.2).

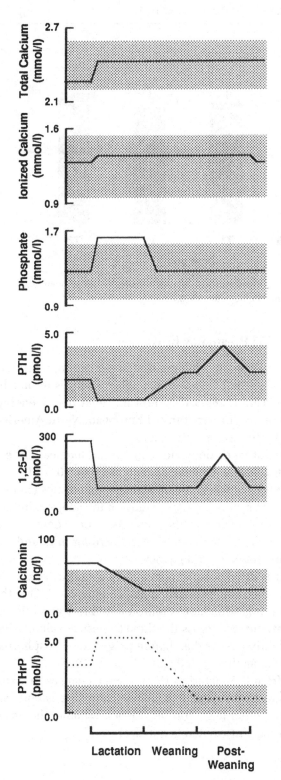

**Figure 3.12.** Schematic illustration of the longitudinal changes in calcium, phosphate, and calcitropic hormone levels in women that occur during lactation and after weaning. Normal adult ranges are indicated by the shaded areas. The progression in PTHrP levels has been depicted by a dashed line to indicate that the data are less certain. (From C.S. Kovacs and H.M. Kronenberg, "Maternal-fetal calcium and bone metabolism during pregnancy, puerperium, and lactation," *Endocrine Reviews* 18:832–72, 1997, © The Endocrine Society, used with permission.)

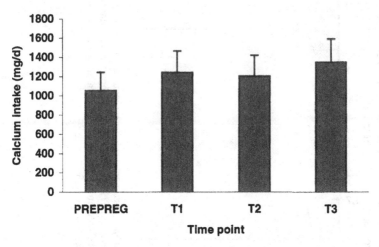

**Figure 3.13.** Daily calcium ingestions by pregnant women before pregnancy and during the trimesters. (Adapted from Ritchie et al., 1998.)

## Calcium Intake During Pregnancy in Women

A number of reported ingestive changes that occur during reproduction in women (Prentice et al., 1993) may perhaps reflect changes in metabolic requirements (Prentice et al., 1996a,b; Ritchie et al., 1998). In fact, energy intake was increased quite clearly in one study of European, North American, and Australian women (Prentice et al., 1996a,b).

There is some suggestive evidence for calcium preference. For example, one study of cravings during pregnancy reported that 50% of women increased their milk intake in the first half of their pregnancies (Hook, 1978), as discussed in the next section. Whereas most of the subjects in that study said they did so out of concern for their infants or on a doctor's advice, a substantial number reported having a greater desire for milk and that milk tasted better to them than before pregnancy.

Calcium ingestion is altered during pregnancy (Prentice et al., 1993; Institute of Medicine, 1997; Ritchie et al., 1998) (Figure 3.13). One thorough study tracing calcium ingestion across reproduction in women in northern California found that, indeed, across the three trimesters calcium ingestion per day was significantly greater than before pregnancy. Phosphorus intake was also increased during that period (Ritchie et al., 1998).

We do not know from those studies if there were changes in taste acuity with regard to the perception of calcium. We do know that calcium salts can taste chalky, salty, and bitter (e.g., Tordoff, 1996c). Whether that is accentuated or blunted for pregnant women is not clear.

## Cravings and Aversions During Pregnancy

In humans, pregnancy brings about dietary changes (e.g., Schwab and Axelson, 1984). Many women report new food cravings and aversions that coincide with the onset of pregnancy and disappear postpartum. It may be that cravings are linked to physiological regulation, such as neuropeptide levels in regions of the brain stem that underlie emesis (Carpeneter et al., 1983, 1984). Nausea and vomiting are common during the first trimester of pregnancy (Walker et al., 1985). Aversions to some foods may be reflections of that.

There is no consistent pattern of preferences and aversions, especially not in comparisons across cultures. Some common substances do crop up frequently, such as fruit and fruit juices. Some 50% of women experience cravings during pregnancy. While some studies show that pregnant women prefer sweet tastes (Hook, 1978; Pope et al., 1992), others do not (Dippel and Elias, 1980; Wijewardene et al., 1994).

As indicated earlier, a study looking at dietary changes among 250 pregnant women in New York found milk consumption to be significantly increased (50% increased their milk intake, about 20% out of concern for their infants, and about 12% because they craved milk or because it tasted better than before pregnancy) (Hook, 1978). Some 18% craved ice cream, and 12% craved fruit or fruit juices.

A study of Sri Lankan women, however, found that 65% of those who had cravings craved sour foods, 47% craved meat and fish, and 40% craved unripe fruits (Wijewardene et al., 1994). Only 30% craved ripe fruit, and only 16% craved sweets. That study also looked at some socioeconomic factors and found that women who had married partners of their own choosing had more cravings than did those whose marriages had been arranged.

A study of South African women of various ethnic descents provided an interesting comparison of ethnic and cultural differences (Walker et al., 1985). Fruits were frequently craved by blacks and by women of mixed race, but less so by white and Indian women. White women predominantly craved sweets, and Indian women mostly craved sour foods. In a study of 97 pregnant white adolescents from eastern Tennessee, 86% reported cravings at some point during pregnancy, particularly sweets (53%), chocolate (30%), and fruit (30%) (Pope et al., 1992); 19% craved pickles, and 15% craved ice cream.

In terms of aversion, 23% of the Sri Lankan women reported specific food aversions, 77% of them to rice, and 62% to sweets (Wijewardene et al., 1994). In the study of South African women, all five ethnic groups ranked meat as the number-one aversion, in contrast to the Sri Lankan women (Walker et al., 1985). Among white women, coffee and tea were the next most aversive substances, but none of the other groups reported those items among their most common aversions. Among the New York women, 28% decreased their coffee intake (most in response to nausea), 16% reported an aversion to meat, 9% to Italian sauces, 9% to vegetables, and 8% to poultry (Hook, 1978). Among the Tennessee adolescents, 66% reported aversions that appeared for the first time during pregnancy, primarily to meat, eggs, and pizza (Pope et al., 1992).

Thus there are great variations in food craving and aversion during pregnancy. One study quite explicitly documented the food cravings of four ethnic groups living in Long Beach, California, showing considerable differences in their food preferences (Coronios-Vargas et al., 1992). The findings from all the studies cited suggest that cravings and aversions during pregnancy reflect, to some extent, the culture in which one lives. And such results are consistent with observations of regional differences in beliefs about nutrition during reproduction (e.g., Carruth and Skinner, 1991).

It is not clear that cravings during pregnancy are linked to needs for particular substances or groups of substances. Nor, for that matter, is it clear that specific cravings in humans reflect metabolic needs (Weingarten and Elston, 1990). There are findings indicating that pregnant women prefer salted over non-salted peanuts, chips, and crackers, in contrast to their previous non-pregnant state (Skinner et al., 1998). For example, one study measured pregnant women's food preferences throughout each trimester of pregnancy and postpartum (Bowen, 1992). That study found that women in the third trimester consumed slightly more salty foods than did the others, and those in the third trimester and postpartum rated salty foods as tasting less salty than did women at other stages of pregnancy.

In other studies, researchers have compared pregnant and non-pregnant women to determine their ability to discriminate among different types of solutions, as well as to determine their preferences (Brown and Toma, 1986; Niegowska and Barylko-Pikleina, 1998; Duffy et al., 1998). The pregnant women, for example, were less sensitive to different NaCl concentrations and more strongly preferred salt solutions, as compared with non-pregnant women. Those researchers suspected that the increased ingestion of sodium during pregnancy was related to changes in the hedonic preference for salt.

## Pica and Pregnancy

Pica, the practice of eating inappropriate substances (clay, dirt, chalk, etc.), is a phenomenon that has been reported during pregnancy and lactation. As indicated in Chapter 2, some specific pica behaviors may result from the

**Table 3.3.** *Substances That Have
Been Known to Be Craved*

Baking powder
Baking soda
Chalk
Cigarette ashes
Cleanser, powdered
Cornstarch
Cough drops
Detergent, powdered
Dirt
Flour
Freezer frost
Ice, crushed, cubed, chopped
Match tips
Milk, dry powdered
Powder, baby and adult
Snow
Toothpaste

*Source:* Adapted from Cooksey (1995).

body's greater demands for calcium. During pregnancy, the extent to which pica is linked to greater needs for calcium, iron, or other solutes is not clear.

Pica can be an adaptation (the search for and ingestion of new substances following physiological changes) or an aberration (the ingestion of substances that are potentially harmful). In the search to satisfy cravings and explore new food sources during pregnancy, women are at increased risk of ingesting something harmful. Assessment of pica during pregnancy must take account of both forms. The range of substances craved is illustrated in Table 3.3.

Case reports of pica are fairly commonplace (Hoener et al., 1991; Barton et al., 1992). In a study some years ago, about 55% of pregnant women in rural Georgia admitted to eating clay. A more recent study described the pica behavior of some relatives of those rural Georgia women who were pregnant and were living in Washington, D.C., and found no incidence of clay or dirt pica, suggesting that the practice was left behind subsequent to northern migration and generational differences (Edwards et al., 1994).

In some cases, pica during pregnancy has been linked to iron deficiency. Iron deficiency has long been known during pregnancy (Stepto and Keith, 1971; Lackey, 1983) and can result in increased salt ingestion (Shapiro and Linas, 1985).

The phenomenon of ice consumption has been noted since antiquity (Parry-Jones, 1992). In a study of pregnant women in Washington, D.C., 8% reported eating large amounts of ice or freezer frost, and a smaller percentage reported eating starch (Edwards et al., 1994). Those women had significantly lower serum ferritin concentrations during the second and third trimesters than did women who did not engage in pica. Women with pica had less iron and calcium in their diets than those without pica.

The consumption of starch during pregnancy is another fairly common pregnancy-related pica practice. In a study of such behavior (known as amylophagia) among urban American women, anemia was twice as frequent among the starch-eaters as among the non-starch-eaters, although it appeared to have no adverse effects on labor (Institute of Medicine, 1997). Further, the rate of pre-term births among amylophagics (measured as infant birth weight less than 2500 g) was almost two times that for the other women. However, 1-minute Apgar scores for infants born to the two groups were not significantly different.

A case report of one pregnant woman's experience with baking-powder pica suggests a link to increased blood pressure, as well as liver dysfunction (Barton et al., 1992). Among other ingredients, baking powder contains calcium acid phosphate and calcium sulfate.

## Calcium Ingestion and Lactation in Women

Calcium ingestion increases during lactation, though not always reaching statistical significance (cf. Insititute of Medicine, 1997; Laskey et al., 1998; Ritchie et al., 1998). In a study in Great Britain, the average intake of calcium was significantly greater for lactating women than for formula feeders or for non-pregnant and non-lactating women (Figure 3.14).

Ann Prentice and her colleagues (e.g., Prentice et al., 1983; Prentice and Barclay, 1991) conducted important studies of differential calcium ingestion cross-culturally during pregnancy and lactation, comparing intakes of calcium by women in Gambia and Zaire and women in Great Britain. They found that calcium intake was considerably lower for the Gambian women than for those from Great Britain (Table 3.4), but so was the ingestion of other substances. The lower levels of calcium ingestion were reflected in lower levels of calcium in breast milk, but calcium supplementation had no effect on milk calcium (Prentice et al., 1983; Prentice and Barclay, 1991), and there were no differences in bone-mineral content between the two groups. The recommended calcium ingestion by country is shown in Table 3.5.

Our attitudes about food choices and our beliefs about nutrition during pregnancy depend to some extent on cultural influences (Carruth and

**Table 3.4.** *Dietary Intakes of Gambian and British Lactating Mothers at 13 Weeks of Lactation*

|  | Gambian Women | British Women |
|---|---|---|
| Calcium (mmol/day) | 7.20 ± 3.20 | 29.20 ± 7.90 |
| Phosphorus (mmol/day) | 26.7 ± 8.3 | 47.9 ± 14.5 |
| Energy (MJ/day) | 7.78 ± 2.06 | 10.2 ± 2.62 |
| Protein (g/day) | 59.2 ± 16.6 | 81.4 ± 26.5 |
| Fat (g/day) | 51.8 ± 24.4 | 100.3 ± 37.3 |

**Table 3.5.** *Recommended Dietary Intake of Calcium in Several Countries (milligrams/day)*

| Country | Non-pregnant Women | Pregnant Women | Lactating Women |
|---|---|---|---|
| Australia | 800 | 1100 (+300) | 1200 (+400) |
| FAO/WHO[a] | 450 | 1100 (+650) | 1100 (+650) |
| France | 800 | 1000 (+200) | 1200 (+400) |
| Indonesia | 500 | 600 (+100) | 600 (+100) |
| Ireland | 800 | 1200 (400) | 1200 (+400) |
| Spain | 600 | 1325 (+725) | 1425 (+825) |
| United Kingdom | 700 | 700 (0) | 1250 (+550) |
| United States | 800 | 1200 (+400) | 1200 (+400) |

[a] Food and Agriculture Organization of the United Nations and World Health Organization recommendations for worldwide average.

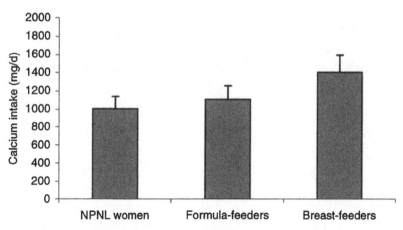

**Figure 3.14.** Daily calcium ingestions by non-pregnant non-lactating women (NPNL), lactating formula-feeders, and lactating breast-feeding women. (Adapted from Laskey et al., 1998.)

Skinner, 1991). That should not be that surprising; we are social animals. The question is to what extent a specific hunger for calcium emerges within the confines of that cultural milieu.

## Summary

There has been selective pressure to develop and use behavioral mechanisms, in addition to physiological mechanisms, in the regulation of calcium that is linked to reproductive fitness. The gonadal steroid hormones, at least in rats, determine the ingestion of calcium. Moreover, in animal models, calcium ingestion increases during pregnancy and lactation, and in humans there are alterations in taste acuity and food preferences during pregnancy and lactation. Calcium intake also increases in humans during pregnancy and lactation. Some of these effects on ingestive behavior may reflect levels of $1,25\text{-}(OH)_2\text{-}D_3$ or estrogen mediated via changes in neuropeptides in the brain (see Chapter 4).

Calcium ingestion is culturally distinct. Culture has an impact on ingestion, such that people in some cultures ingest much more calcium than people in other cultures. Whereas calcium ingestion and calcium hunger have been fairly well delineated in lower species, in women it is less clear. Pica seems to reflect a nonspecific ingestive response that often occurred in the past and in some instances still occurs. The ingestive patterns in humans appear to be strongly influenced by the cultural milieu.

## CHAPTER FOUR

# Neural Endocrine Regulation of Calcium Ingestion

This chapter will describe the hormonal, physiological, and brain mechanisms that appear to underlie cravings for calcium and other minerals. The emphasis in this chapter is on specific hormonal systems, their expressions in the brain, their effects on calcium-binding proteins, and more generally their effects on calcium ingestion. One hypothesis is that a neural endocrine circuit evolved to handle sodium appetite and other regulatory behaviors and perhaps has been co-opted to handle calcium appetite. In other words, a single neural circuit may underlie a number of hormone-regulated behaviors, including those to satisfy mineral appetites (Schulkin, 1991c). First some background on calcium regulation.

## Hormones of Calcium Homeostasis

Vitamin D [as 1,25-$(OH)_2$-$D_3$], parathyroid hormone (PTH), and calcitonin are the most important hormones involved in regulating calcium.

### 1,25-$(OH)_2$-$D_3$

Vitamin D is produced in the skin and converted to the active metabolite [1,25-$(OH)_2$-$D_3$] by the liver and the kidney. Vitamin D is not, after all, a vitamin, but a hormone. A vitamin is defined partly by the fact that the body cannot produce it. In fact, the body does produce vitamin D, and its structure and constituents are steroid in form and function (DeLuca, 1985; Holick, 1994).

Vitamin D is part of a larger family of hormones that include the gonadal and adrenal steroids, in addition to thyroid hormone, all linked by a receptor superfamily (Evans, 1988; Evans and Arriza, 1989). Like other steroids, vitamin D is derived from cholesterol. The chemical structure, synthesis, and metabolic pathways are depicted in Figure 4.1 (DeLuca, 1985). It is initially induced in the skin by ultraviolet radiation. That sets off a cascade of events to form the active metabolite, which then affects end organs in the maintenance of calcium and phosphate regulation.

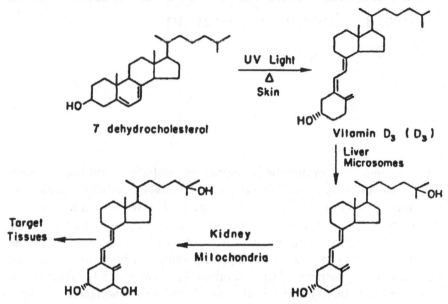

**Figure 4.1.** Physiological metabolism of vitamin D to its in vivo metabolites. (Adapted from DeLuca, 1985.)

Traditionally, steroids were thought to have only genomic effects (e.g., McEwen, 1995). Typically, those effects were believed to take hours to evoke gene products. However, we now know that steroids have membrane-related effects that can be rapid. 1,25-$(OH)_2$-$D_3$ has both genomic and membrane-related effects on tissue (McEwen, 1995).

1,25-$(OH)_2$-$D_3$ binds to intracellular receptors (DeLuca, 1985; Evans, 1988) that consist of ligand-dependent transcription factors that bind to DNA sequences. There are three forms of the receptors, each with a molecular mass of about 35,000 daltons (Darwish and DeLuca, 1996a; Jehan and DeLuca, 1997) (Figure 4.2).

Vitamin D works in three ways to elevate calcium and phosphorus in plasma. First, it stimulates the small intestine to transport calcium and phosphorus from the lumen to the plasma. Second, it mobilizes calcium from the skeleton with the help of PTH. Third, it again combines with PTH to activate reabsorption of calcium by the distal renal tubule (DeLuca, 1988; Institute of Medicine, 1997).

## PTH

PTH is not produced exclusively in the parathyroid glands, because it is also found in the brain and in the placenta. This hormone consists of 84 amino

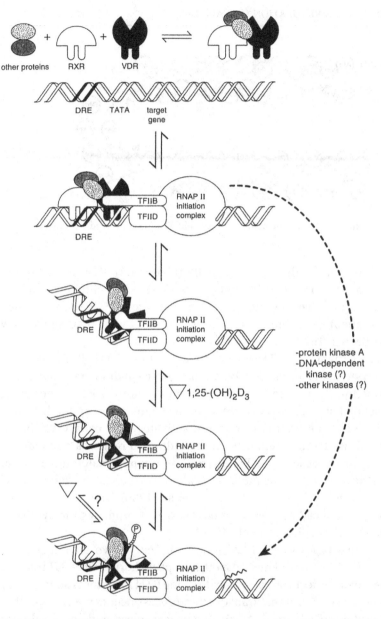

**Figure 4.2.** A model for vitamin-D-mediated gene expression. This model shows the vitamin D receptor (VDR) bound to the vitamin D response element (DRE) complexed with the retinoic acid X receptor (RXR). The RXR binds to the 5'-half site of the DRE, and the VDR binds to the 3'-half site. The RXR-VDR heterodimer binds the VDR in the absence of 1,25-$(OH)_2$-$D_3$. Other factor(s) seem to be required for the complex formation with DREs, but their identity is not yet known. The VDR-RXR complex interacts with the transcription machinery through direct interaction between the VDR and the transcription factor TFIIB. The binding of 1,25-$(OH)_2$-$D_3$ to the VDR induces receptor phosphorylation. The phosphorylated receptor bound to ligand and to the other proteins stimulates transcription. (Adapted from Darwish and DeLuca, 1996a.)

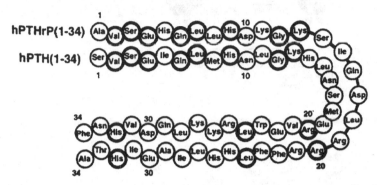

**Figure 4.3.** Amino-terminal amino acid sequence of hPTHrP is compared with that of hPTH.

acids and has a molecular weight of 9500 (Figure 4.3). PTH is stored and released from secretory granules that are synthesized by chief cells, which largely make up the parathyroid gland. The half-life of PTH when secreted in plasma is about 2–5 minutes before it is removed by the liver and kidney (Segre and Brown, 1996).

PTH-related peptide (PTHrP) is distinct from PTH (Strewler and Nissenson, 1996). It is not responsive to changes in calcium metabolism and thus its function appears to be different from the normal PTH function in calcium metabolism, at least with regard to the regulation of systemic tissue.

A low blood concentration of calcium activates PTH cells via calcium sensors to release PTH, which acts on the kidney to retain calcium and to inhibit phosphate reabsorption. Parathyroid cells contain calcium sensors at the level of the membrane, and binding of calcium to the sensors reduces PTH (Figure 4.4) (Segre and Brown, 1996; Cole et al., 1999). The converse holds true if there is reduced sensory occupancy, which promotes the synthesis and secretion of PTH (Brown et al., 1998).

There are two types of PTH receptors coupled to two signaling pathways linked to G proteins. Thus, like other peptide hormones, PTH induces membrane-related effects on receptor sites. One PTH receptor is activated by both PTH and PTHrP, and that transduction process occurs via changes that involve G proteins (Usdin et al., 1996), adenylate cyclase, and phospholipase C as second messengers (Usdin et al., 1996). There is a third receptor that appears to be specific for PTH and is responsive to C-terminal PTH fragments.

## Calcitonin

Calcitonin is a 32-amino-acid peptide hormone produced in the thyroid gland (and the brain). In the thyroid gland it is released in response to

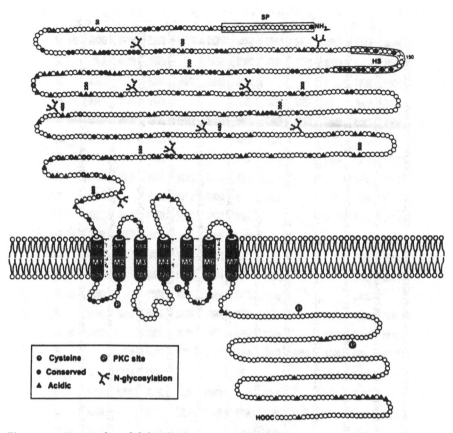

**Figure 4.4.** Proposed model for the bovine parathyroid calcium-sensing receptor protein, which includes seven membrane-spanning helices. The large amino-terminal region is located extracellularly and contains nine potential glycosylation sites. Amino acids are indicated at intervals of 50 in this region. Potential intracellular protein kinase C phosphorylation sites are shown. Amino acids that are identical in the bovine parathyroid calcium-sensing receptors are indicated. (From Segre and Brown, 1996, with permission.)

hypercalcemia and is inhibited by hypocalcemia. For example, high rates of calcium ingestion and elevated levels of calcium in plasma facilitate calcitonin secretion (Raynaud et al., 1994; Deftos, 1996). Calcitonin inhibits bone resorption and may promote removal of phosphate from the plasma.

Calcitonin is a diverse peptide in its variability of expression in different species. There is also a calcitonin-gene-related peptide that has 37 amino acids (Raynaud et al., 1994; Deftos, 1996) (Figure 4.5).

Elevated levels of calcium promote the synthesis and systemic circulation of calcitonin (Deftos, 1996). Circulating concentrations of calcitonin in humans range from 0.1 to 1.0 mU. Calcitonin stimulates cyclic adenosine

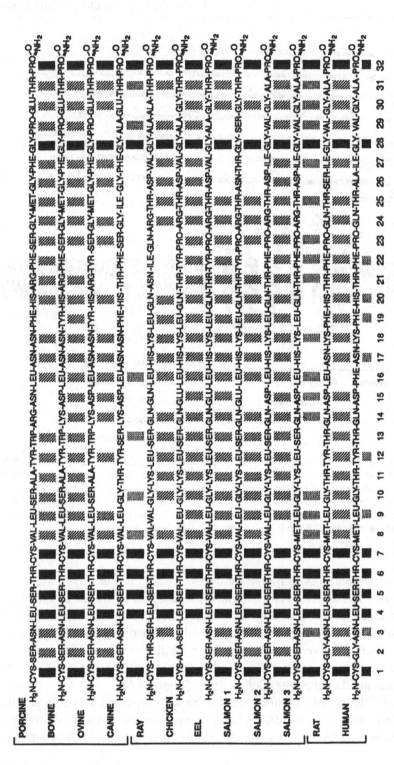

Figure 4.5. Amino acid structure of calcitonin.

monophosphate (cAMP). Both calcitonin and calcitonin-related peptide (mainly produced in the brain) are products of the calcitonin gene (Deftos, 1996; Institute of Medicine, 1997).

The release of calcitonin, like that of PTH, is controlled by two intracellular mechanisms. The first is calcium, and the second is cAMP. Calcium levels in the extracellular fluids also affect calcitonin synthesis and secretion. However, $1,25\text{-}(OH)_2\text{-}D_3$, not the calcium level, controls the inhibition of calcitonin-gene expression per se (DeLuca, 1992; Institute of Medicine, 1997).

## Secretion of Calcium Hormones Following Calcium Deficiency

Tordoff and associates reported a study (Figure 4.6) in which rats were deprived of calcium for periods ranging from 3 days to 3 weeks, and then the circulating concentrations of PTH, calcitonin, and $1,25\text{-}(OH)_2\text{-}D_3$, the active metabolite, were measured. In calcium-deprived rats, plasma concentrations of ionized calcium, bound calcium, and total calcium were reduced. Plasma levels of PTH and $1,25\text{-}(OH)_2\text{-}D_3$ were significantly elevated. Calcitonin levels (not shown in the figure) were initially reduced, but after 14 days they were similar to those in controls (Tordoff et al., 1998). Calcium deprivation raised the plasma PTH concentration from 25 to 50 pg/ml and the $1,25\text{-}(OH)_2\text{-}D_3$ concentration from 29 to 92 pg/ml. Calcitonin decreased from 32 to 25 pg/ml, then increased to 39 pg/ml. Those authors cautioned that the calcium deprivation was very mild and that the rats used were fairly large and had considerable calcium stores in bone on which to draw (Tordoff et al., 1998).

## Calcium Hormones and Calcium Appetite

Richter and Birmingham (1941) demonstrated that parathyroidectomized rats' intake of calcium could be reduced either by implanting PTH in the eyes or by administering vitamin D. The hormones presumably restored normal retention and distribution.

## Vitamin D and the Brain

The nuclear localizations of the active metabolite $1,25\text{-}(OH)_2\text{-}D_3$ are widely distributed in the body and the brain (Stumpf et al., 1979; Stumpf and O'Brien, 1987; Luine et al., 1987; Stumpf, 1995) (Figure 4.7). In localization experiments, animals were treated with radiolabeled $1,25\text{-}(OH)_2\text{-}D_3$. Receptors for $1,25\text{-}(OH)_2\text{-}D_3$ have been found in turtles, rodents, fish, and birds (Stumpf, 1995), and the distributions of the receptor sites throughout the

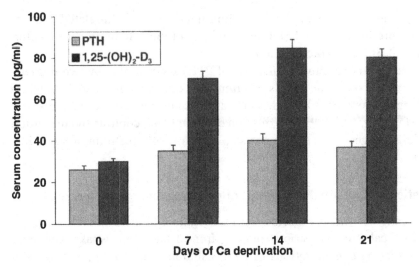

**Figure 4.6.** Elevations of vitamin D and parathyroid hormone following calcium deficiency. Plasma hormone concentrations in rats fed a calcium-deficient diet for various periods. (Adapted from Tordoff et al., 1998.)

body are similar among species. They range from neocortical, pallocortical, and hypothalamus to the brain stem and spinal chord (Stumpf and O'Brien, 1987; Luine et al., 1987). The receptors are also found in the nucleus accombens, the substantia innominata, the preoptic and septal regions, and the ventral hippocampus, as well as in midbrain structures such as the dorsal raphe nucleus, the parabrachial nuclei, and cranial sensory nuclei of the trigeminal nerve.

Importantly, regulation of $1,25\text{-}(OH)_2\text{-}D_3$ receptors has been demonstrated in sites in the brain that are known to regulate appetite, including the paraventricular nucleus and ventral medial nucleus of the hypothalamus, the medial and central nuclei of the amygdala, and the medial bed nucleus of the stria terminalis (Stumpf et al., 1979; Stumpf, 1995) (Figure 4.8).

## $1,25\text{-}(OH)_2\text{-}D_3$ Effects on Calcium Ingestion

Could $1,25\text{-}(OH)_2\text{-}D_3$ generate a calcium appetite? The idea is simple: The same hormone that conserves calcium might also act in the brain to influence the search for and ingestion of calcium – the hypothesis that the active metabolite of vitamin D might contribute toward eliciting calcium appetite (Leshem et al., 1996). To test that idea, rats were injected with different doses of $1,25\text{-}(OH)_2\text{-}D_3$ (25 ng, 75 ng, or 225 ng) for 1 week in a counterbalanced design. Calcium ingestion was measured 24 hours later and was found to have occurred, except with the 25-ng dose. Water intake was also

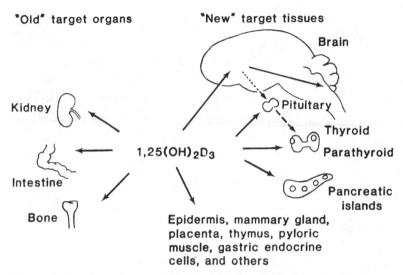

"Old" target organs     "New" target tissues

Brain

Kidney

Pituitary

Thyroid

1,25(OH)$_2$D$_3$    Parathyroid

Intestine

Pancreatic islands

Bone

Epidermis, mammary gland, placenta, thymus, pyloric muscle, gastric endocrine cells, and others

**Figure 4.7.** Target cells for 1,25-(OH)$_2$-D$_3$ (solid lines) have been identified by autoradiography, not only in the three classic target organs (left), but also in many new target tissues (right). Evidence for a vitamin-D-related pituitary-thyroid functional connection (dashed line) has been reported, and target cells for 1,25-(OH)$_2$-D$_3$ are found in both neurons and pituitary cells. Therefore, a related brain-pituitary link is postulated (dotted line), and the existence of a neuroendocrine brain-pituitary-intestinal axis, involved in central regulation of calcium homeostasis, is proposed. Like other steroid hormones, 1,25-(OH)$_2$-D$_3$ may act on many tissues, some of which do not seem to be directly related to calcium metabolism, but involve other functions as well. (Reprinted with permission from W.E. Stumpf, M. Sar, S.A. Clark, and H.F. DeLuca, "Brain target sites for 1,25-dihydroxyvitamin D$_3$," *Science* 215:1403–5, 1982. Copyright 1982 American Association for the Advancement of Science.)

increased. However, the findings were less clear in subsequent replications (Figure 4.9).

In a much more extensive study (Tordoff et al., 1998) of the effects of 1,25-(OH)$_2$-D$_3$ infusion on calcium ingestion by calcium-deficient rats, the rats were fed a calcium-deficient diet and offered a 50-mM solution of CaCl$_2$ as their only source of calcium. They were infused with various amounts of 1,25-(OH)$_2$-D$_3$ for 13 days. Blood samples were taken to analyze total calcium in plasma using colorimetry. The researchers estimated the amount of protein-bound calcium by subtracting the amount of ionized calcium from the total calcium concentration. They also measured hematocrit, sodium and potassium concentrations, and femur morphology, among other things.

Infusions of 1,25-(OH)$_2$-D$_3$ ranging from 2 ng/hour to 16 ng/hour elicited increases in CaCl$_2$ intake in the first 2 days of the experiment (Tordoff et al., 1998). After the first 2 days, the rats that received moderate infusions (2–4 ng/hour) of the metabolite continued to ingest more CaCl$_2$ than did controls throughout the experiment, although the effects decreased as the infusions

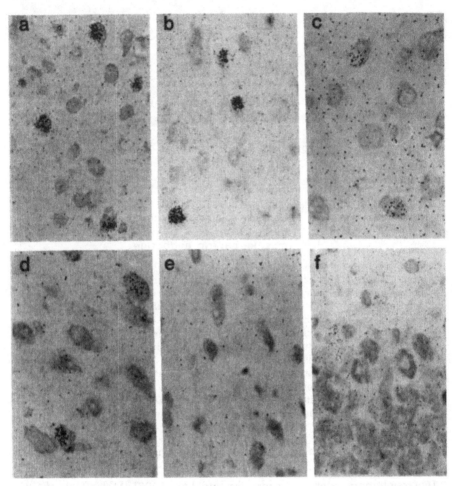

**Figure 4.8.** Autoradiograms of rat brain showing nuclear concentration of radioactivity 2 hours after injection of $^3$H-labeled 1,25-(OH)$_2$-D$_3$ in neurons of (a) the nucleus centralis of the amygdala, (b) the nucleus interstitialis of the stria terminalis, and (c) the outer zone of the nucleus spinalis caudalis of the trigeminus. In competition studies designed to chemically characterize the nuclear radioactivity, unlabeled substances were injected before treatment with labeled 1,25-(OH)$_2$-D$_3$; (d) 25-(OH)-D (nucleus centralis amygdalae) did not prevent nuclear uptake; (e) 1,25-(OH)$_2$-D$_3$ (nucleus interstitialis striae terminalis) prevented nuclear uptake of radioactivity; (f) neurons of the cerebellar cortex do not show nuclear labeling in the same animals that show labeling in parts a–d. Stain, methyl green pyronine. Magnification (a,b,e, and f), ×560; (c and d), ×880. Exposure time (a–c, e, and f), 150 days; (d) 75 days. (Reprinted with permission from W.E. Stumpf, M. Sar, S.A. Clark, and H.F. DeLuca, "Brain target sites for 1,25-dihydroxyvitamin D$_3$," *Science* 215: 1403–5, 1982. Copyright 1982 American Association for the Advancement of Science.)

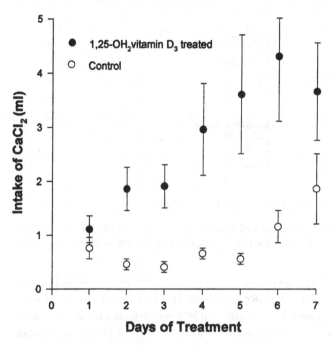

**Figure 4.9.** Effects of injections of vitamin D on 24-hour calcium ingestion by rats. (Adapted from Leshem et al., 1996.)

progressed. The initial increases in calcium ingestion among those rats that received the highest doses of 1,25-(OH)$_2$-D$_3$ (8 ng/h and 16 ng/h) were followed by pronounced decreases in calcium ingestion on days 3–8 of the experiment. Figure 4.10 shows mean alterations in calcium ingestion over a 10-day period.

Those findings, taken together with the known distribution of 1,25-(OH)$_2$-D$_3$ in regions of the brain that regulate other mineral appetites, suggest a behavioral role for 1,25-(OH)$_2$-D$_3$: that of eliciting calcium appetite to defend calcium homeostasis. But consider that in response to the 1,25-(OH)$_2$-D$_3$ infusions, the rats also demonstrated dose-related increases in plasma calcium concentrations. Such findings suggest that calcium concentrations in plasma may underlie the regulation of the appetite (Tordoff et al., 1998). In addition, the infusions led to dose-related increases in plasma calcitonin concentrations and decreases in plasma PTH concentrations, consistent with a rise in serum ionized calcium.

How is it that both elevation and deficiency of vitamin D can result in calcium ingestion? It is possible that the peptide PTH expressed in the brain (as discussed later) plays a role in calcium ingestion similar to the role that

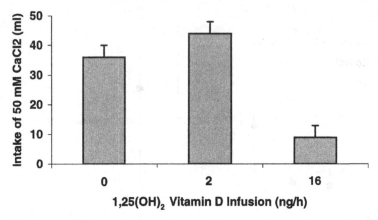

**Figure 4.10.** Effect of chronic infusions of vitamin D on 24-hour intake of 50-mM CaCl$_2$ (collapsed over a 10-day period) by intact rats. (Adapted from Tordoff and Okiyama, 1996.)

angiotensin in the brain plays in sodium appetite (Fitzsimons, 1979; Epstein, 1982; Denton, 1982). It is also possible that calcium-binding proteins are responsible for the increased ingestion; calcium-binding proteins are known to be altered in animals with excessive levels of vitamin D and in animals with deficient levels, as discussed later. These are speculative hypotheses that require further investigation.

## PTH and PTHrP and Receptors in the Brain

PTH is secreted in the periphery during calcium deficiency (as noted in Chapter 3). Because it is a peptide hormone with between 35 and 84 amino acids, it does not readily cross the blood-brain barrier. However, it does have access to the brain, as do other peptides, via circumventricular organs.

These are brain regions outside the blood-brain barrier. cDNA coding within the brain demonstrates that PTH is produced in the brain (Nutley et al., 1995), but far less is produced, for example, in the hypothalamus than in the parathyroid gland (Nutley et al., 1995). Parathyroid-hormone-like immunoreactive cell bodies have been found in the pituitary gland and in regions of the hypothalamus in bullfrog, goldfish, and sheep brains (Balabanova et al., 1986; Kaneko and Pang, 1987; Pang et al., 1988) (Figure 4.11).

PTH and PTHrP receptor sites have been localized in a wide array of brain regions (Weir et al., 1990; Harvey and Hayer, 1993; Weaver et al., 1995b) (Figure 4.12) using in situ hybridization histochemistry. They include regions of the hypothalamus, striatum, hippocampus, bed nucleus of the stria terminalis, amygdala, and neocortex. The receptor sites include the midline

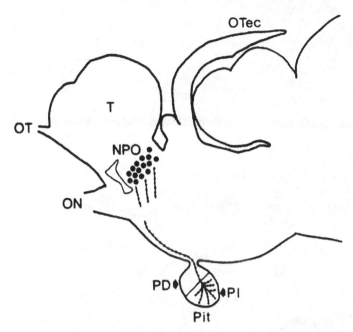

**Figure 4.11.** Schematic drawing of sagittal section of the goldfish forebrain and pituitary gland showing distribution of immunoreactive cell bodies (closed circles) and fibers (dotted lines). NPO, nucleus preoptic; ON, optic nerve; OT, olfactory tract; OTec, optic tectum; PD, pars distalis; PI, pars intermedia; Pit, pituitary gland; T, telencephalon. (From Kaneko and Pang, 1987, with permission.)

thalamic nuclei and pontine nuclei. As in the periphery, there also is a distinction between PTH-gene expression in the brain and PTHrP. They seem to be biochemically dissociated; their functional differences are less clear (Weir et al., 1990).

There is even the suggestion that PTH in the brain may participate in the regulation of calcium balance in the systemic circulation (Matsui et al., 1995). For example, PTH infusion into the cerebral ventricles elevates plasma levels of calcium (Matsui et al., 1995). PTH infused into the lateral ventricle also alters the firing patterns of neurons within the ventral medial hypothalamus, one region linked to the regulation of appetite (Matusi et al., 1995).

## PTH Effect on Calcium Ingestion

In a study with broiler chickens (Lobaugh et al., 1981) it was observed that calcium-deprived chickens, which normally ingested calcium when offered

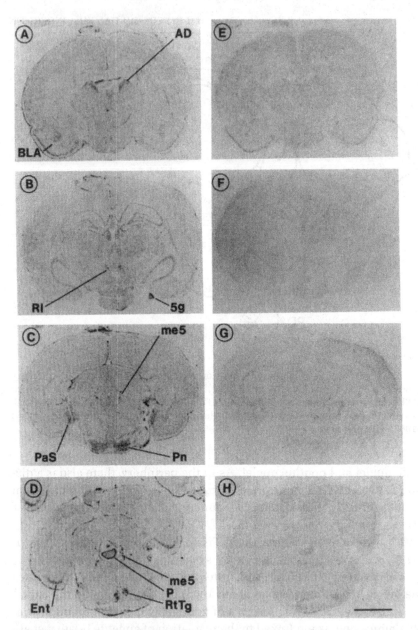

**Figure 4.12.** Distribution of PTH/PTHrP gene expression in adult rat brain at rostral levels. Hybridization signal appears black. The images in panels A–D were generated from sections hybridized with the antisense probe. AD, anterodorsal nucleus, thalamus; BLA, basolateral amygdaloid nucleus; Ent, entorhinal cortex; me5, mesencephalic nucleus, trigeminal; P, Purkinje-cell layer of cerebellum; PaS, parasubiculum; Pn, pontine nucleus; RI, rostral interstitial nucleus, medial longitudinal fasciculus; RtTg, reticulotegmental nucleus; 5g, trigeminal ganglion. (Adapted from *Molecular Brain Research* 28, D.R. Weaver, J.D. Deeds, K. Lee, and G.V. Serge, "Localization of parathyroid hormone-related peptide (PTHrP) and PTH/PTHrP receptor mRNAs in rat brain," pp. 296–310, copyright 1995, with permission from Elsevier Science.)

it, decreased their calcium ingestion when infused intravenously with PTH (one dose, 60 U/kg). Interestingly, in that study, calcium infusion also decreased their calcium ingestion.

In studies of the effects of systemic PTH infusion in calcium-deprived rats (Tordoff et al., 1998), the rats were given thyroid-parathyroidectomy and implanted with pumps that released varying doses of PTH, ranging from none (vehicle only) to 160 ng/hour. When the parathyroid gland was removed, rats craved calcium. When infused with low levels (40 ng/hour) of PTH, their appetites returned to normal. When given higher doses, their appetites for calcium were reduced below baseline levels.

It should be noted that in that study, PTH was infused systemically and not directly into the brain. PTH, as noted earlier, does not cross the blood-brain barrier. It is still possible that injections of PTH centrally might actually increase calcium ingestion and might be further augmented by $1,25\text{-}(OH)_2\text{-}D_3$. But that is speculation, and it is based on the model of sodium appetite, where central injection of angiotensin increases sodium appetite, and a background of aldosterone potentiates that effect (e.g., Epstein, 1982; Fluharty and Sakai, 1995).

Thus, the theory that the hormones of calcium metabolism are produced in the periphery (in this case PTH, because $1,25\text{-}(OH)_2\text{-}D_3$ is not produced in the brain) would seem to suggest that PTH does not play a causal role in the appetite. However, because PTH is itself expressed in the brain, and the brain is the engine of behavior, that still leaves open the possibility that changes in PTH-gene expression in the brain are induced by calcium deficiency, by vitamin D or $1,25\text{-}(OH)_2\text{-}D_3$ deficiency, or by whatever means may be part of the genesis of calcium appetite. This hypothesis requires further investigation.

## Calcitonin and the Brain

Calcitonin, like PTH, is a peptide hormone that does not cross the blood-brain barrier, but is produced in the brain. Calcitonin is a 32-amino-acid peptide hormone secreted by the thyroid gland in mammals (Skofitsch and Jacobowitz, 1992).

Calcitonin is widely distributed in the brain in a wide variety of species that have been studied, with both receptor sites and the peptide itself being synthesized in a wide array of brain regions (MacInnes et al., 1982; Hilton et al., 1995): the cortex; the nucleus accumbens; the organum vasculosum of the lamina terminalis; the caudate putamen; the arcuate nucleus; regions of the amygdala, including the lateral nucleus; brain-stem sites, including the pontine nuclei, the area postrema, and the substantia nigra; the nucleus of the solitary tract; the hypoglossal region; the lateral septum; and the bed nucleus of the stria terminalis. Calcitonin has been studied in numerous species, including humans. There are different forms of calcitonin peptide or related peptide and receptor sites in the brain. One such distribution in the forebrain is depicted in Figure 4.13.

It is not known if calcitonin-gene expression, at the level of cell or receptor body, is altered by calcium deficiency or by changes in the hormones of calcium homeostasis.

## Calcitonin Effects on Ingestion

It has been suggested that calcitonin might act to inhibit the intake of calcium, because it is known that calcitonin counteracts the actions of PTH and $1,25\text{-}(OH)_2\text{-}D_3$. Moreover, one consistent effect of calcitonin is its inhibition of food intake in both animals and people (Chait et al., 1995). In studies in which calcitonin was injected directly into different sites in the brain, the paraventricular nucleus of the hypothalamus was the most sensitive for inhibiting food intake following infusions. That is not surprising, given our prior understanding of the role of the paraventricular nucleus of the hypothalamus in regulating food intake (Dallman et al., 1995). In a study in humans, peripheral infusion of calcitonin or cholecystokinin decreases food intake (Chait et al., 1995). On the basis of such studies, it is not clear to what extent calcitonin-induced inhibition of food intake is due to general malaise or to an actual satiety signal. Decreased food intake obviously leads to a decrease in mineral intake, including calcium and sodium.

In the study cited earlier involving calcium-deprived broiler chickens, intravenous infusions of calcitonin had no effect on calcium ingestion (Lobaugh et al., 1981). Another study of thyroid-parathyroidectomized (TPTX) and thyroidectomized (THX) calcium-deprived rats evaluated the effects of systemic calcitonin infusion (Tordoff et al., 1998). Infusions ranged from 4 to 32 ng/hour in the TPTX rats and 8–64 ng/hour in the THX rats. The higher infusions (at least 16 ng/hour in TPTX rats and at least 32 ng/hour in THX rats) increased $CaCl_2$ intake for 2–6 days and were associated with transient decreases in total plasma calcium concentrations. Those researchers hypothesized that calcitonin's effect on plasma calcium concentration (but not elevated calcitonin levels themselves) was responsible for the increase in $CaCl_2$ intake (Tordoff et al., 1998). Whereas calcitonin infused peripherally is not thought to cross the blood-brain barrier, perhaps central infusions of calcitonin might decrease calcium appetite.

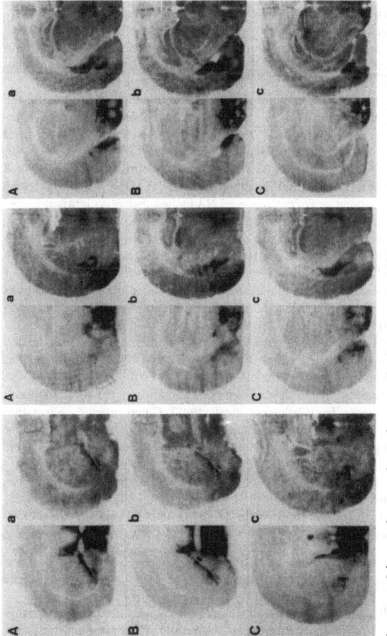

**Figure 4.13.** Schematic drawings of coronal sections of rat brain according to the atlas of Paxinos and Watson (1982). Coordinates are given in millimeters anterior or posterior to the interaural plane. The density and distribution of calcitonin-binding sites are indicated in panels A–C, and calcitonin-gene-related-peptide-binding sites are indicated in panels a–c. Semiquantitative ratings (low, moderate, dense, and very dense) are indicated with different shades. (Adapted from Skofitsch and Jacobowitz, 1992.)

## Calcium Transport and Calcium-Binding Proteins
## and Calcium Appetite

Perhaps one of the ways in which calcium appetite may be generated is via changes in calcium-binding proteins or calcium receptors in the brain. With regard to calcium-binding proteins, they are widely distributed in the brain (e.g., Winsky et al., 1989, 1992; Jacobowitz and Winsky, 1991) and are regulated by hormones such as $1,25\text{-}(OH)_2\text{-}D_3$ (Sonnenberg et al., 1984).

Calretinin is a calcium-binding protein that is widely distributed in the brain. Its distribution is in regions that underlie behavioral responses, including the medial amygdala, the anterior hypothalamus, and the ventral tegmental area of the brain stem (Jacobowitz and Winsky, 1991).

In one experiment, young rats placed on a calcium-deficient diet for 3 weeks demonstrated a calcium appetite and expressed a substantial (28%) decrease in alterations of calretinin mRNA in the substantia nigra compacta–ventral tegmental area (Strauss et al., 1994). That region of the brain innervates the nucleus accumbens, regions of the amygdala, and the bed nucleus of the stria terminalis (Alheid et al., 1996) and therefore is anatomically positioned to influence the search for calcium when it is needed.

In another experiment, glucocorticoids were found to regulate calretinin in regions of the brain that underlie mineral appetite (medial nucleus of the amygdala) (Strauss et al., 1995). When rats were treated with corticosterone over a 7-day period, there were substantial reductions of calretinin mRNA in the paraventricular nucleus of the hypothalamus (93%) and the medial amygdala and the nucleus reuniens (40%).

Calbindin is regulated by corticosterone and appears to act intraneurally as a calcium buffer (Christakos et al., 1987). Glucocorticoids affect that calcium-binding protein in the hippocampal neurons; adrenalectomy reduces the protein, and glucocorticoid reinstates it (Iacopino and Christakos, 1990).

Finally, there are several types of calcium-binding proteins that are dependent on $1,25\text{-}(OH)_2\text{-}D_3$ for their expression in the brain; $1,25\text{-}(OH)_2\text{-}D_3$ affects the calcium-binding protein D-CaBP in both the brain and the periphery (Sonnenberg et al., 1984).

## Calcium Receptors

Two mechanisms are essential in maintaining calcium homeostasis: sensors that detect changes in calcium concentration and effector systems that transport calcium between extracellular spaces (Hebert and Brown, 1995; Hebert et al., 1997). Binding of calcium-sensing receptors activates phospholipase C. That leads to increased levels of diacylglycerol and inositol and the release of calcium, which is then sustained by the influx of $Ca^{2+}$ through membrane-related changes.

Moreover, there are calcium receptors in the central nervous system as well as the periphery. Using in situ hybridization histochemistry, calcium receptors originally found in the kidney have been mapped in the brain, and they appear to be related to a glutamate receptor linked to a G protein (Rogers et al., 1997). Regions of the brain with high concentrations include the olfactory bulb, hippocampus, and circumventricular organs such as the subfornical organ. They also include the medial preoptic and ventral medial hypothalamus, amygdala, bed nucleus of the stria terminalis, zona incerta, and many regions of the hypothalamus (Rogers et al., 1997) (Figure 4.14). It is not known if calcium deprivation results in changes in those calcium receptors.

## Stress Hormones and Calcium Ingestion

Stress is known to increase sodium ingestion in a variety of species. For example, the stress caused by being restrained will increase sodium ingestion; the hormones of stress – corticotropin-releasing hormone (CRH), adrenocorticotropic hormone (ACTH), and cortisol – are elevated, and each of these is known to influence salt consumption (Denton, 1982; Tarjan and Denton, 1991).

Central CRH and ACTH infusions elicit increased sodium and water ingestion (Weisinger et al., 1978; Tarjan and Denton, 1991; Blair-West et al., 1996). It is not known if CRH affects calcium intake. In addition, ACTH systemic injections in several species increase both water and sodium consumption (Blaine et al., 1975; Weisinger et al., 1978). Research on rats and rabbits has shown that the increased appetite is specific to sodium, and intakes of calcium and other minerals are not changed (Weisinger et al., 1978; Denton, 1982). However, when oxytocin is combined with ACTH, calcium ingestion is increased (Denton, 1982). Finally, cortisol injections in some instances increase the ingestion of calcium in rabbits, in addition to ingestions of sodium and potassium (Denton, 1982).

Perhaps the mineral ingestion that occurs when the hormones of stress are elevated reflects the function of the central nucleus of the amygdala and the lateral bed nucleus of the stria terminalis. The central nucleus of the amygdala is a region that underlies gustation as well as cardiovascular and hemodynamic homeostasis (Johnson and Thunhorst, 1997); the same holds true for the bed nucleus of the stria terminalis. Perhaps most important is

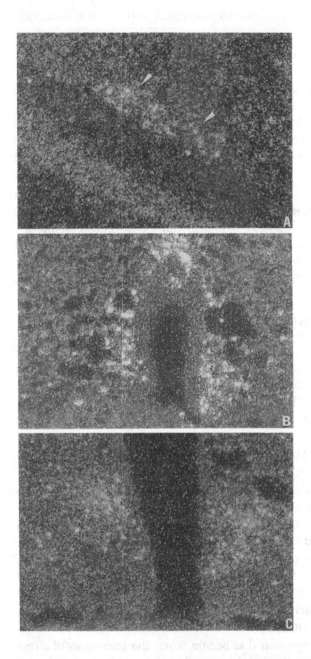

**Figure 4.14.** In situ localization of calcium-receptor (CaR) mRNA in the diencephalon. Darkfield photomicrographs of antisense CaR probe hybridization in (top) zona incerta, (middle) anterior medial preoptic nucleus, and (bottom) dorsomedial part of the ventromedial nucleus. Magnification, ×25. (Reprinted from *Brain Research* 774, K.V. Rogers, C.K. Dunn, S.C. Hebert, and E.M. Brown, "Localization of calcium receptor mRNA in the adult rat central nervous system by in situ hybridization," pp. 47–56, copyright 1997, with permission from Elsevier Science.)

the fact that glucocorticoids increase the expression of CRH in those two regions (e.g., Swanson and Simmons, 1989; Watts and Sanchez-Watts, 1995), which may result in the increased desire for water, sodium, and, perhaps, calcium.

## Calcium Deficiency, Stress Hormones, and Salt Appetite in Rats

In an interesting study, Tordoff (1996a) studied the relationships among adrenalectomy and replacement of aldosterone and sodium and calcium ingestions. Rats were on calcium-deficient diets of varying magnitudes. Typically, adrenalectomy elicits sodium ingestion (Richter, 1936). But the adrenalectomy that increased sodium ingestion by normal rats decreased sodium ingestion by calcium-deficient rats (Tordoff, 1996a). Aldosterone replacement did not alter that behavior. Tordoff (1996a) suggested that corticosterone might be playing a role in facilitating the appetite for sodium in calcium-deficient rats.

In another study, rats placed on a calcium-deficient diet were assessed for the timing of their ingestions of sodium salts and the levels of stress hormones that were circulating (Tordoff and Okiyama, 1996). Importantly, on the endocrine side, both corticosterone and ACTH levels were elevated. Not surprisingly, rats ingested the sodium during their dark cycle; they are nocturnal animals, and that is the time when they would be foraging for food (Rowland et al., 1985; Rosenwasser and Adler, 1986).

It is also known that corticosterone potentiates calcium-deficiency-induced sodium ingestion (Tordoff, 1996a). Corticosterone treatment facilitates a number of behaviors, including mineralocorticoid-induced sodium appetite (e.g., Ma et al., 1993), central angiotensin-induced thirst (Sumners et al., 1991), and CRH-induced fear (Schulkin, 1999). As the molecules of energy homeostasis (Dallman et al., 1992, 1995; Sapolsky, 1992; McEwen, 1998), glucocorticoids sustain ongoing behaviors, in this case salt ingestion.

## Sodium and Calcium Hormones and Sodium and Calcium Ingestion

The hormones of sodium homeostasis, angiotensin and aldosterone, have not been linked to calcium ingestion. Nonetheless, calcium deprivation can alter the circulating levels of these hormones. Angiotensin is a hormone that is produced in the periphery and through enzymatic processes becomes the active metabolite angiotensin II, which mediates many of the effects (Fitzsimons, 1998). Aldosterone is produced in the adrenal gland

and contains receptors called type-I-preferring corticosteroid receptor sites (McEwen, 1995). Both hormones act in the periphery to conserve and redistribute sodium and act in the brain to generate sodium appetite (Epstein, 1982; Fluharty and Sakai, 1995).

Interestingly, polyethylene glycol (PEG), which activates the renin-angiotensin and aldosterone systems and is a well-known elicitor of sodium ingestion (Fitzsimons, 1979), also elicits calcium ingestion (Tordoff, 1997b). A recent study found that PEG also depleted calcium from extracellular fluid and created a calcium appetite (Tordoff, 1997b). There was no change in plasma total calcium concentrations, but plasma ionized-calcium concentrations were reduced.

As shown in Chapter 2, calcium deprivation induces a sodium appetite. However, the hormones of sodium homeostasis, when blocked, do not interfere with calcium-deprivation-induced sodium ingestion (Tordoff, 1996b). In other words, when central blockade of angiotensin is produced by pharmacological agents, there is no effect on the sodium ingestion that results from calcium deficiency. This is in marked contrast to the inhibition of sodium intake that results from sodium depletion (Fluharty and Sakai, 1995). Similarly, there is no effect on the sodium ingestion that occurs following calcium deprivation when there is pharmacological treatment with aldosterone- or angiotensin-receptor blockers (Tordoff, 1996b). This is in marked contrast to the inhibition that occurs in sodium-depleted rats (Wolf, 1969b; Sakai et al., 1989).

However, there is an amazing effect that occurs in calcium-deprived rats ingesting sodium: Rats on a calcium-deficient diet were given access to sodium salts or other solutions to ingest. Those rats that were given the sodium or calcium solutions to drink had higher plasma calcium concentrations and reduced PTH and $1,25\text{-}(OH)_2\text{-}D_3$ levels in their plasma. That did not occur in the rats given saccharin or glucose to ingest (Tordoff, 1997a). Those findings suggest that the NaCl ingestion that can occur from calcium deficiency (Schulkin, 1981; Tordoff et al., 1990) can, for a brief time at least, have an effect on whole-body physiology in the regulation of calcium (Figure 4.15).

The problem with this, however, is that it provides a false sense of relief from the point of view of the body. It reminds one of the ingestion of saccharin. Rats ingest saccharin in part because it is sweet (e.g., Zucker, 1969), and from a biological point of view it should signal energy. But it does not provide any nourishment. What is interesting is how profound a response the saltiness of the sodium solution is for the calcium-hungry rat. However, from a long-term view, increased sodium intake can lead to increased urinary calcium loss and a worsening of calcium deficiency.

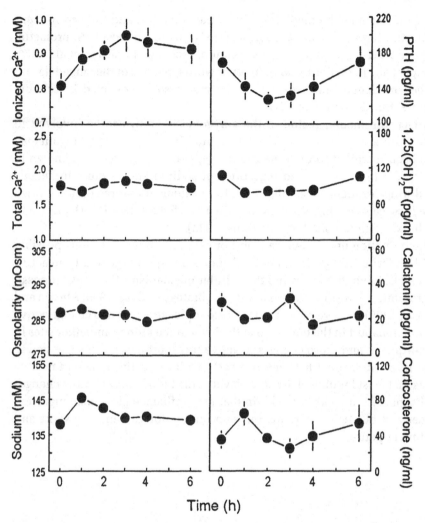

**Figure 4.15.** Data on plasma minerals, hormones, and other factors collected from rats fed a calcium-deficient diet and given 300-mM NaCl to drink for 20 minutes, starting at time 0. Vertical lines show SE; SE was smaller than the symbol size where these are not shown. (From Tordoff, 1997a, with permission.)

## Summary of the Brain Regions Underlying Mineral Appetite

The search for calcium and the ingestion of calcium are mediated in part by neuroendocrine events. Hormones linked to calcium ingestion act to facilitate and sustain central states, perhaps through changes in calcium transport and calcium-binding proteins and activation of neuropeptides (e.g., prolactin) or neurotransmitters.

Regions such as the medial region of the amygdala and the bed nucleus of the stria terminalis contain $1,25\text{-}(OH)_2\text{-}D_3$, estrogen receptors, prolactin, PTH and PTHrP receptor sites, and calcium receptors (McEwen et al., 1977; Stumpf et al., 1979; Pfaff, 1980). In other words, regions of the amygdala and the bed nucleus of the stria terminalis may also be involved in calcium appetite. But this is speculative.

In fact, the medial region of the amygdala is known to be involved in sodium appetite (Nitabach et al., 1989; Schulkin, 1998), and that region may have been recruited to underlie calcium appetite. That region is known to play a role in the increased ingestions of sodium and calcium by female rats during pregnancy (Schulkin, 1991c). Lesions of that region are known to reduce calcium ingestion by multiparous female rats (H. DeLuca and J. Schulkin, unpublished observations, 1988).

The induction of calcium transport or calcium-binding proteins possibly could underlie changes in appetite. Calcium transport is generally an active process. Calcium transport and the cellular mechanisms that make it possible are similar across crustaceans and vertebrates (Simkiss, 1996; Ahearn and Zhuang, 1996). Mineralocorticoids, for example, are known to change active sodium transport in the brain and at the level of the kidney and other sites in the periphery, and those changes in active transport have been suggested to underlie sodium appetite (Denton, 1982). Similarly, active transport mechanisms that exist for phosphate at the level of the kidney may parallel changes in phosphate transport in the brain that may influence phosphate ingestion (Sweeny et al., 1998). Perhaps the same holds for calcium transport and calcium appetite.

## CHAPTER FIVE

# Nutritional Issues, Bodily Health, and Calcium Regulation over the Life Cycle

In this chapter, we begin with a discussion of current nutritional recommendations for calcium, vitamin D, and phosphorus ingestion. We then turn to the hormonal interactions between calcium ingestion and maintenance of bone metabolism and health. That is followed by a discussion of the disorders of calcium metabolism and the beneficial effects of calcium ingestion for human health.

Although the importance of calcium both for general tissue health and as a potential buffer against a number of medical conditions involving bone is increasingly realized, calcium is not a drug; calcium ingestion and its beneficial effects are strongly affected by lifestyles.

Moreover, if most people died in their thirties or fifties, as they once did, low calcium intake would not be an issue. However, the fact that people are continuing to live longer means that calcium plays an increasingly important role in health and disease (Power et al., 1999). For example, not too many years ago, relatively few women underwent menopause, because typically women did not live beyond their childbearing age. Thus menopause and the consequent permanent loss of bone mineralization constitute a modern dilemma.

## Nutritional Recommendations for Calcium

The calcium content of the adult body is 1100–1200 g. Calcium accounts for 1–2% of total adult body weight, and more than 99% of the body's calcium resides in the bones and teeth (Institute of Medicine, 1997).

Requirements and recommendations for dietary intake of calcium vary by age groups, for numerous reasons, including developmental and functional physiological status. We are literally walking around on our calcium reserves. Again it is important to emphasize that while calcium intake is fundamental in maintaining bone health, more than calcium is required. Protein and vitamins C, D, and K are also important, and they can serve as co-factors in deficiency. Several studies have shown that poor calcium intake is associated with reduced intake of other important minerals and vitamins

**Table 5.1.** *Recommended Dietary Calcium Intakes in the United States (mg/day)*

| | |
|---|---|
| 0–6 months | 210 |
| 6–12 months | 270 |
| 1–3 years | 500 |
| 4–8 years | 800 |
| 9–18 years | 1300 |
| 19–50 years | 1000 |
| 51–70 years | 1200 |
| >70 years | 1200 |

(e.g., Barger-Lux and Heaney, 1992, 1994; Barger-Lux et al., 1995). Dietary protein also affects calcium absorption (Kerstetter et al., 1998; Institute of Medicine, 1997). Calcium utilization therefore can be improved within the overall context of food intake. The healthier the diet, the more likely it is that calcium will be ingested, absorbed, and utilized.

Although there is great variation in calcium ingestion worldwide, the health consequences of that variation are uncertain (Prentice et al., 1983; Prentice, 1994b). This may reflect adaptations to low-calcium diets over the short run. It is not clear whether or not it will be significant in the long run, although some studies have demonstrated so indirectly. For example, as mentioned in Chapter 3, studies by Prentice and her colleagues have demonstrated huge disparities between Gambian and British women in calcium ingestion during lactation (Prentice et al., 1983; Prentice, 1994b). One result has been lower levels of bone-mineral content in the offspring of the Gambian women, as compared with the British children, in the first 36-month test period (Prentice et al., 1990).

## Calcium Requirements by Age, Sex, and Reproductive Status

Calcium is necessary at all stages of human development. However, current evidence suggests that the highest demand for calcium is during adolescence, followed by older age (e.g., ≥51 years) (Table 5.1).

Almost one-quarter of adult women in the United States take calcium supplements, compared with 14% of adult men and 7.5% of children (Institute of Medicine, 1997). The average dose is 248 mg/day for women, 160 mg/day for men, and 88 mg/day for children. Average daily calcium intakes are higher among men and women who take daily supplements of any kind, but the differences are statistically significant only for women (Institute of Medicine, 1997). Nonetheless, it has been reported that calcium intake has declined over the past 15–20 years (U.S. Department of Health and Human Services, 1991), at the same time that the recommended daily allowance (RDA) by the Food and Drug Administration has increased.

Calcium absorption varies by age. It peaks in infancy (at about 60%) before declining, and it rises again in early puberty (about 34%) (Abrams et al., 1997; Institute of Medicine, 1997); it is maintained at around 25% in young adults, and then declines with age. In postmenopausal women, there is a more rapid decline in calcium absorption of about 0.21% per year (Bullamore et al., 1970; Heaney, 1992), and the data show continuing decrements with age.

A study in adolescent and adult women by Weaver and colleagues (1996) showed that calcium absorption was 50% higher in the adolescents, despite the fact that there were no differences in plasma calcium, 1,25-$(OH)_2$-$D_3$, and PTH levels. It was concluded in that study that young females absorb more calcium from the intestine and excrete less calcium from the kidney. Bone turnover is also higher in adolescents, which reflects the greater calcium absorption and bone growth (Weaver et al., 1996). There have been reports of positive relationships between calcium intake and bone-mineral density in young women (Ramsdale et al., 1994; Teegarden et al., 1998).

An individual's bone mass does not change greatly once it peaks during development, except during periods following gonadal steroid decrements. It appears that maximizing peak bone mass during development may help to prevent osteoporosis and decrease the risk of bone fractures (Matkovic, 1991). The highest rate of turnover, and the greatest need for calcium, is during development. As indicated earlier, adolescents absorb calcium at a high rate (Matkovic, 1991). Furthermore, it has been observed that radial bone absorption is greatest during the maximal growth period of adolescents (Peacock, 1991), and there is some evidence suggesting that calcium supplements during childhood can positively influence femoral-neck bone strength and therefore bone mass (Ammann et al., 1993). This finding holds tremendous implications for preventive health and for the potential benefit from interventions such as calcium fortification of bread, breakfast cereals, and other common foods.

A series of studies conducted in a population of middle-aged women who were generally in slightly negative calcium balance has reinforced the current RDA for aging women with regard to calcium balance and requirements (Heaney et al., 1975, 1977). Their intake of calcium was calibrated for calcium balance, which normally requires about 1200 mg/day.

When calcium homeostasis was studied in women experiencing menarche and menopause, the differences between the two groups were striking. Total absorption of calcium during menarche was 212 mg/day, whereas for menopausal women it was 42 mg/day. That evidence of an age-related decline in calcium utilization and absorption has major connotations for preventive health.

## Vitamin D and Calcium

Calcium scarcity and vitamin D deficiency are intertwined. Whereas calcium intake declined about 10,000 years ago because of the rise of agriculture and increased reliance on food products such as grains, vitamin D intake at that time should not have changed, because people still were predominantly laboring outdoors. The industrial revolution and the increase in urbanization changed all that and provided the context for decreased vitamin D levels. Beginning in the 1600s, rickets was first noted in northern England and other northern European countries, with the incidence increasing during the industrial revolution and persisting into the twentieth century. Young children were particularly vulnerable.

As early as 1822, one researcher proposed that lack of exposure to sunlight was at the root of the problem, observing that children in urban Warsaw had a much higher incidence of rickets than those living in rural areas outside Warsaw. That was confirmed 70 years later when disease rates among children in Great Britain were compared with rates in underdeveloped countries. Children in the northeastern United States also suffered from the disease in disproportionate numbers (Holick, 1994).

Our understanding of the nutritional link to rickets arose from the folk practice of giving cod liver oil to cure the disease. The liver oil effectively prevented and cured rickets. Early in the twentieth century, it was determined that the vitamin D in liver oil was responsible for curing rickets, as reviewed by Holick (1994). Around the same time, other investigators were successfully treating rickets with artificial-light therapy and with exposure to sunlight (Holick, 1994). That led to the addition of provitamin D to milk, which was then irradiated to activate the vitamin D. Later, scientists learned how to add vitamin D to milk directly. In the United States and other countries, rickets was eradicated (DeLuca, 1985; Holick, 1994). Vitamin D has diverse roles in calcium homeostasis and phosphorus homeostasis (Figure 5.1).

The Institute of Medicine recommends a daily dietary intake of 5 $\mu$g of vitamin D for all persons up to age 50 years, including women who are pregnant or lactating. The recommendation rises to 10 $\mu$g per day for those aged 51–70 years, and 15 $\mu$g per day for those over 70 years of age. The upper limit of intake is 25 $\mu$g per day for infants up to 12 months of age and 50 $\mu$g per day for all persons over 1 year of age (Institute of Medicine, 1997).

Dietary intake of vitamin D is difficult to measure, but one large survey has estimated that median consumption by young women is 2.9 $\mu$g per day,

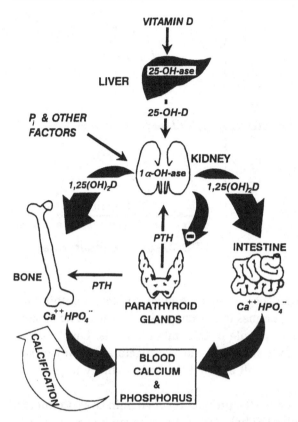

**Figure 5.1.** Metabolism of vitamin D and the biological actions of 1,25-dihydroxy-cholecalciferol [1,25-(OH)$_2$-D$_3$]. (From Holick, 1994, with permission. Copyright Am. J. Clin. Nutr. American Society for Clinical Nutrition.)

versus 2.3 $\mu$g per day by older women (Institute of Medicine, 1997). Among the few foods naturally containing vitamin D are some fish liver oils, flesh of fatty fish, and eggs from hens fed vitamin D (Institute of Medicine, 1997). Most human intake comes from foods fortified with vitamin D, primarily milk and other dairy products. Almost 25% of adults take a vitamin D supplement, which provides an additional 10–20 $\mu$g per day (Institute of Medicine, 1997).

Throughout most of human evolutionary history, however, the primary source of vitamin D was exposure of skin to sunlight. In fact, the longer the light exposure, the greater the vitamin D synthesis in skin. There are seasonal effects, with more vitamin D produced in the spring and summer seasons than in fall and winter months (Holick, 1994) (Figure 5.2). Investigators have evaluated the effects of seasonal change in vitamin D production among residents of Boston, Massachusetts (Holick, 1994). The production of vitamin D

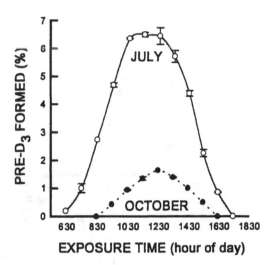

**Figure 5.2.** Photosynthesis of pre-cholecalciferol (previtamin D) at various times on cloudless days in Boston in October and July. (From Holick, 1994, with permission. Copyright Am. J. Clin. Nutr. American Society for Clinical Nutrition.)

in the skin gradually declines, beginning in August, to almost nothing by November. Then production resumes in March, peaking in July. The Boston residents were compared with people in Edmonton, Canada, Los Angeles, California, and San Juan, Puerto Rico, and the investigators concluded that latitude, season, and time of day all affected the body's ability to produce vitamin D.

Sunscreens and clothing reduce the production of vitamin D, as do shields of glass, plastic, and Plexiglas (Holick, 1994). Because such barriers block ultraviolet B radiation, they also block the production of vitamin D in the skin. In older people, whose ability to produce vitamin D is significantly diminished, protection from the sun may lead to further vitamin D deficiency (Holick, 1994).

Thus, as exposure to unfiltered sunlight declined, vitamin D deficiency became commonplace in northern latitudes. That trend, together with decreased calcium consumption, placed a double strain on calcium homeostatic regulation, and today we recognize greater dietary demands for both vitamin D and calcium. The fortification of food, notably dairy products, helps to reduce the threat of calcium deficiency. Nonetheless, populations that ingest small amounts of dairy products may be vulnerable to pathologic abnormalities arising from calcium metabolism.

Another development that has made calcium deficiency an even more significant threat to human health is our longer life expectancy. Low calcium intake, slightly negative calcium balance, elevated levels of PTH, and declining levels of $1,25\text{-}(OH)_2\text{-}D_3$ are all associated with aging and morbidity. When the body is challenged by a low-calcium diet, young adults increase their

1,25-$(OH)_2$-$D_3$ production and intestinal absorption of calcium. However, older adults have impaired responses (Gallagher et al., 1987; DeLuca, 1995). In all species studied, intestinal absorption of calcium diminishes with age.

Apparently, dietary calcium ingested by older people cannot prevent the loss of bone to support the soft-tissue needs for calcium. Serum levels of PTH rise with age (Gallagher et al., 1987; DeLuca, 1995), but in older people the kidney does not effectively produce 1,25-$(OH)_2$-$D_3$ in response to PTH. Adding to these problems, it appears that the numbers of vitamin D receptors diminish with age (Ebeling et al., 1994; DeLuca, 1995).

In countries where calcium intake is high, administration of 1,25-$(OH)_2$-$D_3$ in doses high enough to simulate normal human production increases the risk of hypercalciuria and hypercalcemia. However, it is possible that administering high doses of 1,25-$(OH)_2$-$D_3$ in conjunction with a moderate-calcium diet may be effective in maintaining calcium balance (DeLuca, 1995).

## Phosphorus

Phosphorus regulation is important to consider because it is linked to calcium regulation. Moreover, phosphorus ingestion, like calcium ingestion, varies across cultures (Prentice et al., 1993). As one of those elements vital to developing and maintaining bone, phosphorus makes up about 0.5% of the newborn's body (Institute of Medicine, 1997) and some 0.65–1.1% of the adult body (Aloia et al., 1994; Institute of Medicine, 1997). The greatest concentration of phosphorus is found in bone (85%), and the rest in soft tissue (Institute of Medicine, 1997).

The net dietary absorption of total phosphate ranges between 55% and 70% in adults and 65–90% in infants and children (Institute of Medicine, 1997). This does not appear to vary according to dietary intake. In contrast to the situation with calcium, however, there is no apparent adaptive mechanism that can improve phosphorus absorption at low intakes (Institute of Medicine, 1997). Absorption is compromised by pharmacological doses of calcium carbonate and by antacids that contain aluminum. However, typical intakes of calcium do not appear to affect phosphorus absorption. The utility of determining the appropriate ratio of calcium to phosphate for infants and children has been questioned.

As with calcium, however, inadequate intake of phosphorus can lead to rickets in children and osteomalacia in adults, as well as anorexia, anemia, muscle weakness, bone pain, and other conditions. However, phosphorus is present in so many foods that it takes almost total starvation to cause dietary phosphorus deficiency. The Institute of Medicine recommends a daily intake of 100 mg of phosphorus for infants up to 6 months of age and 275 mg for infants 6–12 months of age. The recommended daily allowance for children aged 1–3 years is 460 mg; at 4–8 years, 500 mg; at 9–18 years, 1250 mg. For adults, the recommended dietary allowance is 700 mg per day. Young pregnant or lactating women (aged 14–18 years) should consume 1250 mg per day, but pregnant and lactating women over age 18 will have sufficient intake with 700 mg per day (Institute of Medicine, 1997). The upper level of intake ranges from 3 to 4 g per day (Institute of Medicine, 1997).

## Calcium Homeostasis and Bone Health Following Pregnancy and Lactation

There appears to be little or no bone loss during pregnancy. However, bone loss does occur during lactation and appears to be unrelated to calcium ingestion (e.g., Prentice, 1994b; Kalkwarf and Specker, 1995; Ritchie et al., 1998; cf. Cross et al., 1995b).

Prentice and her colleagues have suggested that maternal calcium stores seem to be somewhat independent of the amount of calcium that is ingested. They concluded that the increased calcium ingestion that is recommended during pregnancy and lactation may *not* be beneficial, regardless of whether women consume low amounts of calcium (Gambia) or high amounts (Great Britain). It is also unclear that vitamin D supplements are beneficial to lactating mothers (Prentice et al., 1997b).

Prentice's group has shown that adaptation occurs: When there are low levels of calcium in the diet, there is greater variation in intake, and typically there is greater renal sufficiency (less urinary excretion) to optimize the use of those low levels (Prentice et al., 1998). On the other hand, calcium ingestion has been shown to reduce preeclampsia in subsets of high-risk women, as discussed later. The implication is that, as my colleague Mike Power has said on numerous occasions, there may be important differences between short-term adaptations and long-term effects with regard to calcium homeostasis (M. Power, personal communication).

Because breast-feeding results in significant calcium loss for lactating women, researchers have studied whether lactation causes transitory or permanent bone-mineral depletion in women who had breast-fed infants in the past. Their findings have been mixed. For the most part, they suggest that the risk is not significant and that calcium supplements are not beneficial, at least for bone calcium metabolism.

In a small study, healthy women who had recently given birth to normal infants were studied for about 6 months postpartum (Hayslip et al., 1989). Twelve women breast-fed exclusively during the study period, and seven fed their infants with formula (or with breast milk for 1–2 months postpartum and formula for the remaining months). There was an initial evaluation of bone-mineral content by single- and dual-photon absorptiometry at 2 days postpartum, and it showed no significant differences in bone-mineral content between the two groups. At 6 months postpartum, however, the breast-feeding women had significant decreases in lumbar (vertebral) bone-mineral content, decreasing from an average of 1.22 g/cm² at 2 days postpartum to an average of 1.14 g/cm² at 6 months postpartum. The formula-feeding group went from 1.26 g/cm² to 1.24 g/cm², a loss that was not statistically significant. Researchers in that study hypothesized that the bone loss resulted from lactating women's physiological state of increased prolactin production and decreased estrogen production (Hayslip et al., 1989). In that study, lactating women ingested more calcium than did formula-feeding women, but there was no correlation between calcium intake and bone-mineral content. The fact that the bone loss occurred in the lumbar spine, and not in the middle or distal radius, suggests that breast-feeding primarily affects trabecular bone.

Another study looked at the duration of lactation in relation to bone loss (Wardlaw and Pike, 1986). Women who had given birth to and had breast-fed 3 or 4 children were separated into two groups: short-term lactation (less than 6 months) and long-term lactation (more than 6 months). Bone-mineral contents in the midshaft and ultra-distal radius were measured by single-photon absorptiometry; these areas are known to have high levels of either cortical or trabecular bone. In both bone sites, the long-term-lactation group had lower bone-mass measurements than the short-term-lactation group. In the ultra-distal radius, which is made up primarily of trabecular bone, the difference was statistically significant. Those authors also found no correlation between calcium intake and bone loss.

Those same authors later reexamined the association between long-term lactation and trabecular bone loss, extending their study to evaluate the effects on the bone sites most prone to osteoporosis: the lumbar vertebrae and the femoral neck (Koetting and Wardlaw, 1988). The subjects were healthy women who had given birth to up to 4 children and had breast-fed each child for 10 months or more. Those women were compared with nulliparous controls. Bone density in various sites did not differ significantly between the groups, leading those authors to conclude that there was no association between long-term lactation and low bone density.

Other studies have examined whether or not bone-mineral content can be recovered after lactation. Lactating and non-lactating women were evaluated postpartum to determine overall bone-mineral content and bone-mineral density in the lumbar spine and distal radii, as measured using dual-energy x-ray absorptiometry (Kalkwarf and Specker, 1995; Kalkwarf et al., 1997) (Figure 5.3). As in their earlier study, the researchers found that in the first

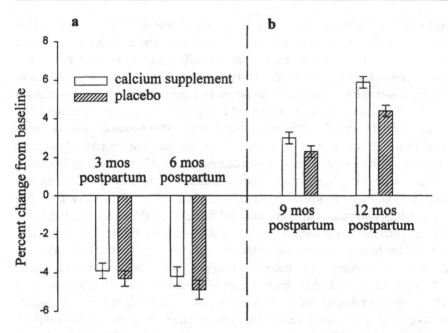

**Figure 5.3.** Effects of lactation, weaning, and calcium supplementation on bone-mineral density in the lumbar spine in two groups of healthy women. (a) Preweaning measurements and changes in bone-mineral density in the lumbar spine in lactating women receiving calcium supplements and placebo. (b) Changes in bone-mineral density in the lumbar spine in women receiving calcium supplements and placebo, measured at approximately 6 months postpartum (pre-weaning) through 12 months postpartum. Weaning occurred, on average, at about 7 months. (Adapted from Kalkwarf et al., 1997.)

6 months postpartum, the lactating women lost more bone overall, and specifically in the lumbar spine, than did the non-lactating women; however, they found no change in the distal radius for either group. After weaning, the lactating women gained more bone than did the non-lactating women. The sooner the lactating women resumed menses, the less bone they lost during lactation, and the more bone they gained after weaning.

Another evaluation of the effects of calcium supplementation on bone density during lactation and following weaning found some change in bone density, but concluded that calcium supplementation did not prevent bone loss (Kalkwarf et al., 1997). That randomized, double-blind, placebo-controlled study evaluated postpartum women with moderate-to-low calcium intake. Lactating women breast-fed their infants for about 6 months; the other group consisted of women who had weaned their babies after breast-feeding almost exclusively for about 6 months. Both groups were paired with non-lactating controls who either had never breast-fed or had done so for 2 weeks or less. Dual-energy x-ray absorptiometry was used to determine bone-mineral content for the total body and bone-mineral densities for the lumbar spine, ultra-distal radius, and distal third of the radius.

The researchers found significant bone loss from the lumbar spine (Kalkwarf et al., 1997). Calcium supplementation had no effect on total-body bone mass or radial-bone density. However, the bone loss was reversed after weaning, and the results suggested that there may be a modest effect of calcium supplements in facilitating such recovery of bone-mineral density in the lumbar spine. In other words, lactation results in bone loss that is inevitable, but then reverses back to stable homeostatic bone balance. Those findings, that lactation does not inhibit recovery from bone loss (Sowers et al., 1993, 1995a,b), have been reported by several additional groups of researchers. Excess calcium ingestion seems not to influence the naturally occurring bone demineralization, nor the restoration process (Sowers et al., 1993; Prentice, 1994b; Kalkwarf et al., 1997).

To summarize, the literature suggests that lactation is a time of increased need for calcium, as expressed in greater bone metabolism (e.g., turnover and resorption). Short-term bone-mineral loss can total 6% over a 6-month period. It seems clear that the bone loss that does occur is restored, with or without future pregnancies. Calcium supplementation is not necessarily beneficial for bone health during pregnancy and lactation (cf. Prentice, 1994b; Kalkwarf et al., 1996; Polatti et al., 1999), but may be of benefit for other tissue, including the fetus and infant, as discussed later.

## Estrogen, Menopause, Calcium Ingestion, and Bone

Calcium and other nutrients are essential for bone growth and influence bone density, but their positive effects on bone should not be considered in isolation from a variety of other factors. Five factors, in particular, are essential for bone density. The first factor to consider is genetics, which helps to determine potential bone health and density. The second factor is physical activity and general health. A third factor is gonadal steroid production, and the remaining two factors are calcium intake and age (Heaney, 1993, 1996).

Consider the normal decreases in bone mass in women (Table 5.2). Bone loss with age, particularly in women in the postmenopausal period of their lives, has been associated with vulnerability to fracture. Estrogen deficiency may be an overriding reason for this bone loss, as estrogen is linked to calcium balance and transport in end-organ sites. Recall the earlier discussion that calcium ingestion during lactation does not restore bone loss, at least not in the amounts that were studied and over the short term (Prentice, 1994b; Kalkwarf et al., 1996, 1997). However, the short-term bone loss that is linked to the absence of estrogen is once again restored with the onset of menses.

**119**

**Table 5.2.** *Decade-Specific Correlation Coefficients Between Bone Mass and Skeletal Size in Women*

| Age Group | Cortical Area D2-d2 on Metacarpal Length | Radius 8 cm BMC/cm and Width | L2-L3-L4 BMC and Body Length |
|---|---|---|---|
| 20–29 | 0.225 | 0.832 | 0.737 |
| 30–39 | 0.537 | 0.566 | 0.658 |
| 40–49 | 0.501 | 0.067 | 0.225 |
| 50–59 | 0.244 | 0.233 | 0.225 |
| 60–69 | 0.413 | 0.260 | 0.094 |
| 70–79 | 0.052 | −0.036 | −0.010 |

The perimenopausal and menopausal periods represent times of major changes in the hormonal milieu governing calcium homeostasis. Estrogen levels begin to decline dramatically, and a number of symptoms begin to emerge: hot flashes, changes in mood, and sleep-pattern disturbances. A decline in estrogen is linked to increased vulnerability to cardiovascular abnormalities, in addition to bone demineralization, as discussed later.

Women lose bone at an accelerated rate, particularly from the spine, for a period of about 5 years after menopause begins (Gallagher et al., 1987; Institute of Medicine, 1997). An increase in calcium during that time does not appear to prevent the bone loss. Parallel with those changes, young women with amenorrhea secondary to anorexia nervosa have high rates of urinary calcium excretion and low rates of bone formation (Institute of Medicine, 1997). Similarly, exercise-induced amenorrhea lowers calcium retention and is associated with diminished bone mass (Drinkwater et al., 1990). But osteoporosis, as a clinical diagnosis, reflects a number of factors besides hypoestrogenemia, including lifestyle, nutrition, health or disease state, and genetics. Decreased estrogen levels are accompanied by decreased calcium absorption efficiency (Gallagher et al., 1980; Heaney, 1996; Institute of Medicine, 1997) and increased rates of bone turnover. In addition, there may be impaired dietary absorption in general.

Menopause, as a predisposing risk factor, results in greater urinary calcium excretion by women not on hormone-replacement therapy (Heaney, 1996). Estrogen is known to have sodium-retaining effects and also seems to have calcium-retaining effects; estrogen facilitates incorporation of calcium into bone, and lack of estrogen results in outflow of calcium from bone (Heaney et al., 1978). Thus, even though calcium requirements are increased, the ability to utilize calcium may be decreased in the first years of menopause in the absence of estrogen.

Bone loss appears to be minimally altered by calcium consumption during the immediate or initial postmenopausal period. Consider one example: In a double-blind, randomized controlled study, one group of women who had recently undergone menopause ingested calcium at less than 400 mg/day, and the remaining subjects ingested more than 1000 mg/day over a 2-year period. Bone-mineral densities in the spine and femoral neck were measured. Women lost bone mass, regardless of calcium supplements (Dawson-Hughes et al., 1990; Dawson-Hughes, 1991).

Thus, bone loss during the perimenopausal period (that surrounding menopause) is reminiscent of that during lactation, when calcium supplementation does not retard bone loss. But in the latter case, bone calcium metabolism is restored with the resumption of menses, whereas the onset of menopause and the loss of cyclical estrogen and progesterone initiate a period after which there is no recovery or restoration of bone loss (Figure 5.4) (Kovacs and Kronenberg, 1997).

Following the initial menopausal period, things begin to look different. A 2-year randomized placebo-controlled trial of calcium supplements (Reid et al., 1995) resulted in fewer fractures and less bone loss. However, the positive effects seemed to decrease over time. In another large study of 77,000 nurses, representing a great age range, there was little evidence of a protective effect from calcium consumption (Feskanich et al., 1997). However, the research findings in this area are equivocal, and in other studies, calcium intake was linked to increases in bone-mineral density (Nieves et al., 1995, 1998).

In several studies, for example, calcium supplements by themselves have been linked to increased bone-mineral density in premenopausal women (Ramsdale et al., 1994). In one study, calcium supplements in premenopausal middle-age women resulted in reduced bone loss postmenopausally, as compared with women who did not take calcium supplements (Smith et al., 1989). In 27 controlled trials of increased calcium ingestion, 26 demonstrated a positive effect in terms of greater bone density and fewer fractures. Among 30 observational studies, 22 of them reported positive effects of calcium intake on bone (R. Heaney, conference remarks, American College of Obstetricians and Gynecologists, 1998).

Estrogen, or hormone-replacement therapy (HRT), can reverse some of the age-related decline in calcium utilization. Estrogen-induced neurotrophic factors in the heart, bone, and brain act to preserve and strengthen cellular maintenance, as reviewed by Plouffe and Schulkin (1998). This does not

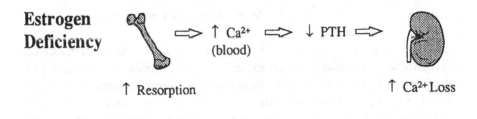

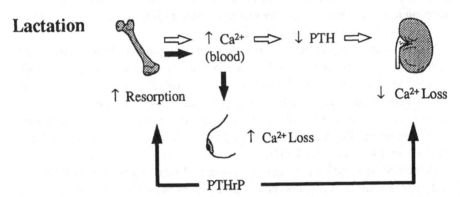

**Figure 5.4.** Schematic illustration of responses to acute estrogen in non-lactating and lactating women. Acute estrogen deficiency increases skeletal resorption and raises the blood calcium; in turn, PTH is suppressed, and renal calcium losses are increased. During lactation, the combined effects of PTHrP and estrogen deficiency increase skeletal resorption and raise the blood calcium, but calcium is directed into breast milk. (Adapted from Kovacs and Kronenberg, 1997.)

mean that HRT is without controversy; published studies have argued for and against a possibly increased risk of cancer. But, on balance, there is more evidence favoring a role for HRT (combination estrogen/progesterone replacement) in women who are not at high risk for ovarian or breast cancer (Baron et al., 1998).

Use of estrogen, coupled with increased calcium ingestion, has been reported to have clear beneficial effects at the level of the calcaneus and the distal and proximal radii (Figure 5.5) (Davis et al., 1995). Replacement estrogen, given with calcium supplements, may slow age-related bone loss (Heaney, 1996). Heaney, in reviewing 31 published studies, suggested that estrogen's protective effect on bone is facilitated by increased calcium intake (1200 mg/day).

In summary, the age-related decrease in estrogen production figures importantly in bone loss in women. Estrogen replacement, coupled with calcium supplements, can have beneficial effects on bone maintenance.

## ADDITIVE EFFECTS OF ESTROGEN AND CALCIUM

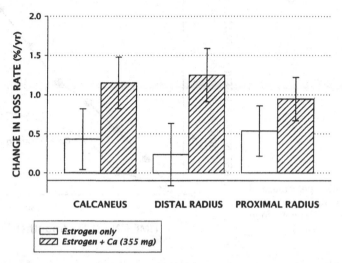

**Figure 5.5.** Decreases in bone loss rate during only estrogen use and when combined with calcium supplement. (Reprinted from *Bone* 17, J.W. Davis, P.D. Ross, N.E. Johnson, and R.D. Wasnich, "Estrogen and calcium supplement use among Japanese-American women: effects upon bone loss when used singly and in combination," pp. 369–73, copyright 1995, with permission from Elsevier Science.)

### Bone, Calcium Ingestion, and Calcium Hormones

Calcium utilization seems to decline over the life cyle, but the mechanisms for that effect are not clear. It may be due to declining levels of $1,25\text{-}(OH)_2\text{-}D_3$ or other calcium-retaining hormones. Vitamin D and exposure to ultraviolet light, as well as renal and fecal functions, decline with age. Serum PTH, by contrast, increases.

Seasonal levels of $1,25\text{-}(OH)_2\text{-}D_3$ are lower in elderly women than in young women (Holick, 1994). HRT is known to effect bone resorption, with the result of lower serum calcium levels, and to increase PTH and $1,25\text{-}(OH)_2\text{-}D_3$ levels in postmenopausal women (Gallagher et al., 1987). There is evidence showing that vitamin D, when coupled with calcium intake, has greater effects on bone metabolism (measured in the spine) than does vitamin D alone (Dawson-Hughes et al., 1997) (Figure 5.6).

Another study in which both groups received $1,25\text{-}(OH)_2\text{-}D_3$ showed that calcium supplementation in conjunction with HRT was more beneficial in reducing bone loss than without HRT, and the subjects receiving HRT had higher levels of total-body calcium (Aloia et al., 1994). Moreover, in a randomized controlled trial of low doses of estrogen and progesterone along

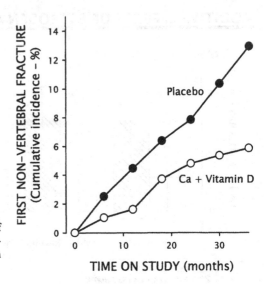

**Figure 5.6.** Cumulative percentage of all 389 subjects with a first non-vertebral fracture. (Adapted from Dawson-Hughes et al., 1997.)

with calcium supplementation and vitamin D, the treatment was quite effective in reducing bone loss in elderly (65-year-old) women (Recker et al., 1999).

It has become increasingly clear that vitamin D deficiency is not uncommon in the elderly (Holick, 1994). This is especially borne out in studies that have compared elderly people confined to indoors and those with access to the outdoors (Gloth et al., 1995). The effects on vitamin D levels are apparent in institutionalized versus non-institutionalized elderly populations (Pietschmann et al., 1990; Fardellone et al., 1995). This is particularly true in the winter (Thomas et al., 1998). In a recent study, low vitamin D intake, reduced sun exposure, winter season, age, and female gender were predictive of vitamin D deficiency in patients.

Vitamin D supplementation is associated with increased bone mass in the femoral neck (e.g., Ooms et al., 1995). However, in one short-term (5-day) trial, $12,5-(OH)_2-D_3$ given orally to women with osteoporosis who were on estrogen replacement therapy did not modify their skeletal responses.

It has been reported that calcitonin also can reduce bone loss in oophorectomized women (Polatti et al., 1993). Moreover, in a 2-year study of calcitonin given to postmenopausal women, it was reported that the calcitonin reduced fractures. There were two study groups, and both were given calcium supplements; in addition, one group was given 100 I.U. of calcitonin for 10 consecutive days each month. The data showed that the calcitonin group had less bone loss and a lower fracture rate (Rico et al., 1992).

Another study showed that PTH given to women on HRT over a 3-year period increased vertebral bone mass. The results for subjects were compared with those for controls on HRT who were not given PTH. Only the PTH group showed increases in vertebral bone mass (Lindsay et al., 1997), as well as decreases in fractures. Those findings are somewhat perplexing in light of evidence that postmenopausal women secrete PTH in a manner similar to younger women (Samuels et al., 1997).

In summary, $1,25\text{-}(OH)_2\text{-}D_3$, and perhaps PTH and estrogen, coupled with calcium ingestion, appears to have a beneficial effect on bone maintenance in postmenopausal women (Recker et al., 1999).

### Physical Activity, Calcium Ingestion, and Bone

Calcium supplements, with or without exercise, reduce the risk of fractures, particularly in women in the highest-risk categories. Some of the factors that contribute to bone fractures are depicted in Figure 5.7 (Heaney, 1996). In addition, there is some (but not much) evidence showing that exercise early in life has a long-term effect on bone mass and strength (Reid, 1996). That sparked the hypothesis that exercise, in addition to increased calcium ingestion (mild powder or tablets), might increase bone density in post-menopausal women (Prince et al., 1995).

We now know that even a high-calcium diet (1000 mg/day), in the absence of physical activity, cannot prevent bone loss. When both physical activity

## DETERMINANTS OF FRAGILITY FRACTURES

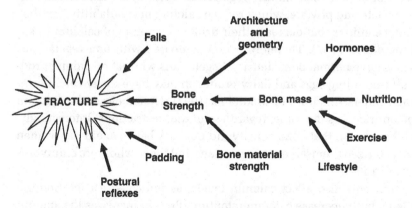

**Figure 5.7.** A comprehensive overview of the osteoporotic fracture context, listing some of the factors that contribute first to fracture, then to skeletal fragility, then to low bone mass, and finally to calcium deficiency (from left to right). (Adapted from Heaney, 1992.)

and calcium intake are increased in children, the rate of bone mineralization increases (Institute of Medicine, 1997). Therefore, the Institute of Medicine does not vary its recommendations for calcium intake in relation to physical activity. Inactivity decreases calcium absorption (LeBlanc et al., 1995), and negative calcium balance can occur following prolonged bed rest, in addition to decreased levels of $1,25\text{-}(OH)_2\text{-}D_3$ and PTH.

Another study looked at exercise and the effects of smoking as factors in maintaining maximal bone mass in young women (Valimaki et al., 1994). Bone-mineral content was up to 10% higher in the group that exercised regularly than in the group having the least amount of exercise. In the same study, women who ingested an average of 1000 mg of calcium daily had approximately 6% greater bone-mineral density in the femoral-neck region than did those who ingested significantly less calcium.

In one study, brisk walking by postmenopausal women was independently associated with lower levels of cortisol and reduced bone turnover when measured up to 72 hours after the exercise (Thorsen et al., 1995). There was no effect on calcitonin and PTH levels.

## Other Dietary Influences on Calcium Ingestion and Absorption

One might assume that certain diets, specifically those of vegetarians and lactose-intolerant people, would affect calcium retention. Lactose-intolerant people are at risk for calcium deficiency because of their decreased intake of dairy products; nonetheless, they absorb calcium from milk normally, as well as from other dietary sources, and so their calcium requirement is no different from that of non-vetgetarians. Vegetarians have higher-than-normal intakes of oxalate and phytate, which reduce calcium bioavailability, but they also produce anions that decrease their urinary excretion of calcium (Institute of Medicine, 1997). Therefore, when compared with non-vegetarians, vegetarians have normal bone density. Vegetarians who eat no animal products at all (including eggs and dairy products) may have lower calcium intakes than others (Institute of Medicine, 1997). Nonetheless, the calcium requirement remains the same, regardless of diet restrictions. Interestingly, research has shown that bone density did not vary between elderly women who characterized themselves as vegetarians and those who were omnivores (Reed et al., 1994).

Other nutrients also affect calcium levels, as indicated earlier. Sodium and caffeine both decrease calcium retention. Protein increases the amount of calcium excreted. However, these effects do not warrant a change in the amount of calcium recommended (Institute of Medicine, 1997).

126

The studies in postmenopausal women, when taken together, suggest that supplementary calcium can reduce hip fractures. However, that effect is augmented when coupled with hormones and exercise. Immediate post-menopausal bone loss over 5 years is not altered by calcium intake. Calcium ingested by itself, above the threshold amount, will not produce more bone (Heaney, 1996). The reason is that other dietary factors influence calcium absorption and utilization. For example, both high protein intake and sodium ingestion promote urinary calcium loss (Barger-Lux et al., 1995), and estrogen promotes calcium absorption and binding. Levels of vitamin D are important in maintaining bone health.

## Calcium and Neural Toxicity

Calcium is an intracellular signal in all cells. The brain is no different and is also the largest consumer of glucose (Sapolsky, 1992). Glucocorticoids, as the hormones of energy balance, promote intracellular changes, and long-term elevation of glucocorticoids may endanger a number of end-target tissues, including brain and bone tissue, by their actions on calcium mobilization. Neural toxicity can occur due to the buildup of intracellular calcium, perhaps mediated by glutamate and N-methyl-D-aspartate receptors that move calcium into the cell (Sapolsky, 1992). One result can be cerebral ischemia (Sapolsky, 1992).

## Disorders of Calcium Regulation

Two types of disturbances in calcium homeostasis are clinically apparent: hypercalcemia and hypocalcemia. The range of serum calcium is depicted in Figure 5.8, which indicates when the patient is hypercalcemic or hypocalcemic (Bushinsky and Monk, 1998).

### *Hypercalcemia*

Hypercalcemia disrupts the normal function of the distal tubule in the kidney, thereby interfering with normal kidney function. This can lead to polyuria, a common occurrence in patients with renal failure. It can occur in patients on vitamin D replacement undergoing dialysis (Bushinsky and Monk, 1998). The patient who is hypercalcemic typically is constipated, has decreased appetite, and may be nauseous. Most patients do not show symptoms until their plasma levels are beyond 12–14 mg/dl. Table 5.3 lists

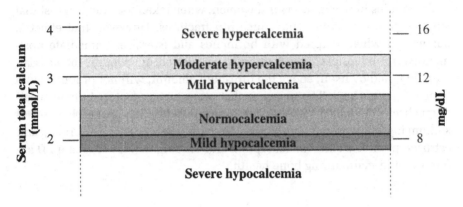

**Figure 5.8.** Ranges of serum calcium during normal and pathological conditions. (Adapted from Bushinsky and Monk, 1998.)

**Table 5.3.** *Common Causes of Hypercalcemia*

| | |
|---|---|
| Hyperparathyroidism | Thyrotoxicosis |
| Malignancy | Vitamin intoxication |
| Thiazide diuretics | Renal failure |
| Immobilization | Renal transplantation |
| Granulomatous disease | Milk-alkali syndrome |
| Familial hypocalciuria | Hypercalcemia |

the features associated with hypercalcemia, which include elevated levels of PTH and possibly vitamin D (Bilezikian, 1992; Bushinsky and Monk, 1998).

Excess PTH is not uncommon, ranging in frequency from 1 in 1000 to 1 in 5000 people (Shane, 1991; Bushinsky and Monk, 1998). The excess PTH results from hypersecretion from the parathyroid gland. Hyperparathyroidism also increases levels of $1,25\text{-}(OH)_2\text{-}D_3$. One result is normal regulation of calcium, and feedback regulation when the hormone is suppressed. Treatment may consist of surgical removal of the parathyroid gland (Bushinsky and Monk, 1998).

Parathyroid tumors and other pathological events can set in motion PTH hypersecretion (Silverberg et al., 1997). Moreover, there appears to be a genetic basis in a subset of hyperparathyroid syndromes that accounts for about 10% of reported cases. Familial hypercalcemia has been linked to a mutation on the calcium-receptor gene (Brown et al., 1998).

**Table 5.4.** *Some Common Causes of Hypocalcemia*

| | |
|---|---|
| Hypoparathyroidism | Parathyroidectomy |
| Pseudohypoparathyroiditis | Hypomagnesemia |
| Malignant disease | Acute pancreatitis |
| Rhabdomyolysis | Septic shock |
| Chronic renal insufficiency | Vitamin D deficiency |

Hypercalcemia is sometimes seen in infants and children and usually is associated with "failure to thrive" (Bushinsky, 1998). The features are irritability, decreased appetite, and gastrointestinal discomfort. Williams syndrome can be associated with hypercalcemia during the first year, and these infants tend to be small at birth, with facial abnormalities. The syndrome usually resolves during the first year.

Medical management of hypercalcemia typically consists of lowering calcium concentrations by promoting calcium excretion in the urine. Pharmacological agents include furosemide and calcitonin; in addition there are procedural approaches (e.g., dialysis, parathyroidectomy) (Bushinsky and Monk, 1998). Pharmacological reduction of PTH secretion via calcium-receptor agonists has also been used to reduce plasma calcium levels. Biophosphonates are common tools for reducing hypercalcemia (Bushinsky and Monk, 1998).

## Hypocalcemia

Hypocalcemia is somewhat common in clinical practice. It can occur because of vitamin D deficiency, liver disease, magnesium deficiency, gastrointestinal surgery, and so forth (Bushinsky and Monk, 1998).

Symptoms associated with hypocalcemia are muscular irritability, decreased sensitivity in the fingers or toes, cramps, seizures, tetany, and movement disorders (Tohme and Bilezikian, 1993; Reber and Heath, 1995) (Table 5.4). Cardiac impairments and dementia can also be associated with hypocalcemia. Hypocalcemia is often associated with prostate and breast cancer and bone disease. Hypoparathyroidism has been associated with hypocalcemia (Bushinsky and Monk, 1998).

Hypocalcemia is also fairly common in neonates. It can appear as early as 5–10 days, when it has been associated with vitamin D deficiency and high phosphate concentrations in the mothers. In older infants (2–4 months) it is associated with low intakes of calcium and vitamin D intake (Specker et al., 1985a; Kovacs and Kronenberg, 1997). Hypocalcemia has been observed in diabetic mothers and preeclamptic patients. Management typically consists of calcium supplements, in addition to vitamin D therapy.

## Hypocalcemia, Hypertension, and Calcium Metabolism During Pregnancy

There is increased calcium resorption from the gastrointestinal tract, and bone metabolism and calcium utilization for the fetus to draw on calcium reserves. Pregnant women have an increased ability to absorb calcium, possibly because of increased production of $1,25\text{-}(OH)_2\text{-}D_3$, which acts directly on the gastrointestinal tract to increase calcium absorption (Seely and Graves, 1993). In pregnancy, the placenta serves as an additional site of 1-hydroxylation, which results in vitamin D levels two to three times higher in pregnant women than in non-pregnant controls. This increase occurs as early as the first trimester (Kovacs and Kronenberg, 1997).

It was in the context of this great utilization and need for calcium that researchers noticed an association between diets low in calcium and increases in blood pressure (Belizan et al., 1991, 1997). By the end of the 1970s there were epidemiological suggestions that calcium intake was related to preeclampsia. For example, McCarron's group, in the first part of the 1980s, showed that serum calcium was higher in patients with normal blood pressure (McCarron, 1983; McCarron and Morris, 1985). The findings from a number of experiments portend a possible role for calcium in reducing pregnancy-related hypertension (Villar et al., 1987; Villar and Repke, 1990; Belizan et al., 1991, 1997) (Figure 5.9).

Studies have documented that pregnant women with preeclampsia are hypocalciuric when compared with normotensive pregnant women (Repke, 1994). Additional studies have shown that women with preeclampsia also have lower vitamin D and ionized calcium levels and higher PTH levels than normotensive pregnant women (Repke, 1994b). It has been suggested that the decreased vitamin D leads to a decrease in the ability to absorb calcium from the gastrointestinal tract. The gastrointestinal tract transfers calcium to the fetus, leading to a fall in ionized calcium and a compensatory increase in PTH.

The systolic pressures in children of mothers who had experienced poor maternal-fetal calcium nutrition were dramatically elevated (Belizan et al., 1991, 1997). One reason, perhaps, that calcium supplements can reduce blood pressure (Belizan et al., 1991, 1997) is that they reduce PTH levels (which are elevated) and intracellular calcium and relax smooth-muscle activity in both nulliparous and multiparous women. They also, perhaps, may be linked to reduction of preterm labor by reducing smooth-muscle contractility (Purwar et al., 1996).

Pregnancy-related hypertension is associated with low birth weight, and there is some evidence that calcium can reduce the risk for both events

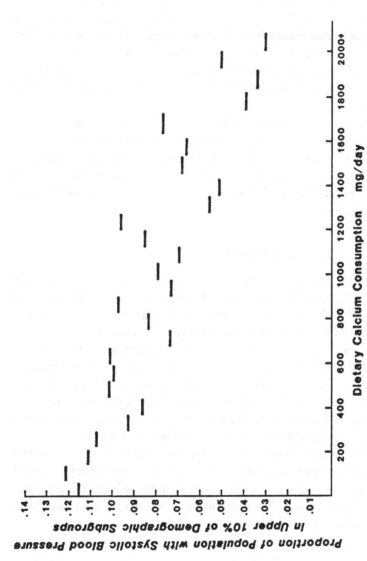

**Figure 5.9.** Percentage of individuals at risk for developing hypertension on the basis of daily calcium intake, as reported in the National Health and Nutrition Examination (NHANES). An apparent threshold intake of 400–600 mg of calcium per day is evident; below that, the risk is increased. Between 600 and 1200 mg the relationship is flat. Above 1200 mg a greater decrease in risk is observed. (From McCarron et al., 1991, with permission. Copyright Am. J. Clin. Nutr. American Society for Clinical Nutrition.)

(Repke and Villar, 1991). In fact, the idea that calcium, added to the normal diet, can reduce hypertension is based on the premise that calcium supplements reduce the expression of hypertension and that calcium deficiency exacerbates that condition. The lower the level of calcium ingestion, the greater the incidence of pregnancy-related hypertension (McCarron, 1998).

The issue is complicated, however. In a CPEP study with more than 4000 women, calcium supplements had no effect, overall, on preeclampsia or pregnancy-related hypertension. Perhaps one way to consider the issue of the links among calcium, hypertension, and pregnancy is in individuals who are vulnerable to hypertension. The effect may be clearer in that group (Dwyer et al., 1998).

For example, one study found that calcium in dairy products was especially helpful in buffering against pregnancy-related hypertension, but also perhaps hypertension in other select populations (e.g., African-Americans) (Dwyer et al., 1998). That dietary approach mixed a low-fat diet (one rich with fruits and vegetables) with the dairy products and resulted in lower levels of hypertension.

## Calcium and Premenstrual Disorder

There are several studies suggesting that calcium balance (Thys-Jacobs et al., 1989, 1995, 1998; Penland and Johnson, 1993), and perhaps PTH and $1,25\text{-}(OH)_2\text{-}D_3$ levels are linked to premenstrual syndrome (Thys-Jacobs et al., 1995, 1998). One controlled study showed that calcium supplements decreased premenstrual syndrome in healthy women aged 18–45 years (Thys-Jacobs et al., 1998). There were daily recordings that included up to 17 features of premenstrual syndrome (e.g., negative affect, bloating, cravings, and pain). Two groups were formed, with one group receiving calcium carbonate at 1200 mg/day, and the other a placebo. The groups were matched for severity of premenstrual symptoms and demographic factors.

During the luteal component of the cycle, the symptoms reported by the placebo group were more severe than those in the calcium-supplemented group. Of the 17 symptoms that were noted, 15 were reduced in the group given calcium supplements. There was up to a 54% decrease in aches and pains (Figure 5.10).

## Calcium, Vitamin D, and Colorectal Cancer

Colon cancer involves a very high mortality as compared with all the various kinds of cancers. Dietary habits have been linked to both colon and rectal

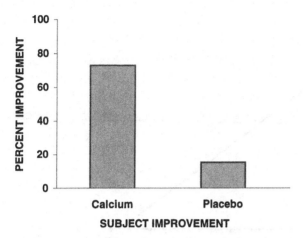

**Figure 5.10.** Calcium supplementation and premenstrual syndrome: percentage of global symptom improvement. (Adapted from Thys-Jacobs et al., 1989.)

cancer (e.g., Garland et al., 1991; Marcus and Newcomb, 1998). In several reports, low levels of calcium consumption have been associated with the greater incidence of colon cancer.

A correlation between calcium ingestion and the risk for colon cancer in both women and men was reported from a study in Utah (Garland et al., 1991). It is clear from that study that the greater the ingestion of calcium, the lower the risk for colon cancer. The findings from several additional studies are depicted in Figure 5.11.

Among the many studies that have been reported, however, there is some inconsistency regarding whether or not there is a strong association between calcium levels and colon cancer. In one study, increased ingestion of calcium was said to be associated with colon cancer. In addition, a reduction in colon cancer has been linked to decreased levels of sunlight and therefore decreased vitamin D levels, along with lower levels of calcium. When taken together, such studies are suggestive that calcium ingestion and absorption may reduce the proliferation of cancerous cells in the colon (Bostick, 1997).

## Summary

The importance of calcium regulation has become well known in the modern age. Longer life and decreases in calcium consumption and photosynthesis of vitamin D have revealed the importance of calcium over the life cyle. But calcium is no miracle drug, and though it is featured high on the top-10 list

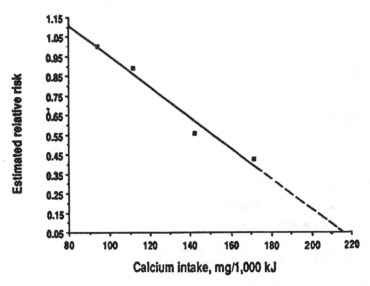

**Figure 5.11.** Estimated relative risk for colon cancer according to calcium intake. (From Garland et al., 1991, with permission. Copyright Am. J. Clin. Nutr. American Society for Clinical Nutrition.)

right now, it is only one important element in the body, along with others. It is a vitally important element, one that we need to maintain for bodily health.

Calcium regulation takes place throughout the whole body and is affected by behavior, kidney functioning, bone health, gastrointestinal-tract status, diet, and exercise. Hormones such as estrogen and vitamin D can reduce the incidence of bone loss and vulnerability to fracture.

There is also, no doubt, an interaction with genetics, one perhaps linked with the clinical manifestations of pregnancy-induced hypertension, premenstrual syndrome, and colon cancer – perhaps reduced by calcium ingestion. Taken as a whole, the regulation of calcium is vital for bodily health.

# Conclusion

There are, as Kanis (1994) has put it, "evangelists who argue endlessly for the importance of high calcium intake and the 'nihilists' who minimize its importance." In fact, Kanis, among others, takes an interesting view of our present predicament. He suggests that human calcium ingestion today is low when compared with that of most other mammals (Eaton and Konner, 1985; Eaton and Nelson, 1991) and that of our ancestors (see Chapter 1). There is reason to believe that our calcium homeostatic physiology evolved under conditions of substantially higher calcium intake than we have today.

Reiterating an earlier theme, the suggestion is that early hominids had a diet that was rich in calcium. Calcium scarcity being less of a problem for them, their priority was perhaps to rid the body of excess calcium. About 10,000 years ago, when agricultural advances appeared, there may have emerged the first context of decreased access to a high-calcium diet. Robert Heaney has depicted the hypothetical paleolithic intake of calcium, the optimal calcium intake now, and minimal daily intakes (Figure C.1).

I have made a point in this book to emphasize that calcium is regulated by both behavior and physiology. In the context of the biological regulation of calcium, behavior is part of the physiological regulation of calcium. In the first chapter, the broad biological context for both types of regulation was presented. The idea of the defense of the internal milieu emerged with Claude Bernard amidst the biological context of adaptation and Darwinian sensibilities. Walter Cannon pushed the concept of the body's defense to maintain homeostasis and internal stability. Curt Richter, the great psychobiologist at Johns Hopkins, developed the idea of behavioral homeostasis.

Calcium ingestion is one example of that paradigm, where behavior and physiology converge in the defense of the internal milieu. I have noted the considerable research in animals and humans on this issue; the literature is replete with examples of increased calcium ingestion in the context of increased need for calcium. This occurs during calcium deficiency or calcium depletion, during pregnancy and lactation, and during development. Aberrations in ingestive behavior, including pica, may also be linked to calcium deficiency, and calcium deficiency may be a factor in lead ingestion.

# THE REQUIREMENT, ADAPTATION, AND AGE

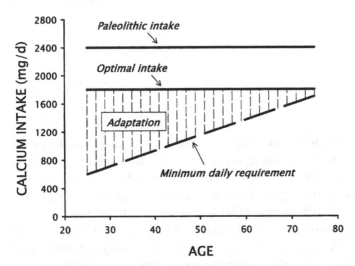

**Figure C.1.** Intake of calcium from an evolutionary and age-related standpoint: optimal and minimal ingestions. (Courtesy of R. Heaney).

There are not many recorded instances of calcium deficiency in humans affecting the ingestion of calcium. There are, however, several instances in which calcium preferences may be related to calcium need. One is in a medical context: dialysis patients, who suffer chronic kidney impairment. Renal failure and dialysis are linked to hypocalcemia, and in one study those patients were offered calcium sources. They are known to have an increased preference for salty foods (Leshem, 1999). These patients are also more likely to prefer calcium-fortified cheese products than are controls (Leshem and Schulkin, 1998) (Figure C.2). Perhaps alterations in PTH or vitamin D levels in patients would result in a similar preference for calcium.

Critical regions of the brain are activated to produce calcium hunger. During times of greater calcium need, calcium transport mechanisms, both at the level of the periphery and in the brain, are altered. The gustatory system appears to be responsive to changes in calcium balance; taste is involved in the regulation of calcium and may be altered by increased demands. The gustatory system traverses from the cranial nerves to forebrain sites. Hedonic sensibilities may underlie the readiness to ingest calcium sources. Moreover, hormones that regulate calcium and are critical to maintaining calcium homeostasis may also contribute to the regulation of appetite.

There is, in fact, wide variation in calcium ingestion across cultures. Recall that in Great Britain, calcium ingestion is quite high. In Gambia, where

**Mean preferred concentration of CaCl in cheese**

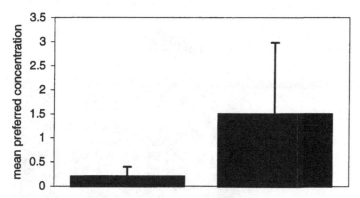

**Figure C.2.** Dialysis patients and mean preferred concentration of CaCl in cheese (controls, left; dialysis patients, right). (M. Leshem and J. Schulkin, unpublished data.)

calcium comes predominantly from plant foods, it is fairly low (Prentice, 1994b). Thus human physiology obviously is capable of adapting to a wide range of calcium intakes. Recall from the Introduction to this book that the range of calcium ingestion is quite striking (Prentice, 1994b). Cultural cuisine (Remington, 1936; Mead, 1943; Rozin, 1976a), social learning (Galef, 1986), and basic gustatory and hedonic responses (Pfaffmann et al., 1977; Stellar and Stellar, 1985) determine most, if not all, of our food choices (Rozin and Schulkin, 1990). In terms of evolution, it has been hypothesized that humans once had access to a calcium-rich diet and were unlikely to have suffered calcium deficits (Eaton and Nelson, 1991; Heaney, 1993).

It is clear that on the clinical side, calcium is critical for a number of end-organ systems and bodily health. Calcium is needed for excitation-contraction coupling (muscle), excitation-secretion coupling (nerve cells, many secreting cells), stability of excitable membranes, cardiac and smooth-muscle action potentials, blood clotting, structural bone, teeth, connective tissue, and ligands for calcium receptors on the parathyroid cell, to name some of the functions of calcium. In addition, new research indicates an important role for calcium not only in bone maintenance but also in premenstrual syndrome, colon cancer, and perhaps some versions of preeclampsia.

The study of calcium homeostasis allows us to assimilate the new findings in molecular biology, melding behavioral and physiological regulation with clinical medicine. Disease states such as hyperparathyroidism are beginning to be studied from this standpoint. Vitamin D, PTH, and calcitonin have been cloned, and their sequences are understood. Moreover, we know molecular similarities and ancestor genes for whole classes of hormones (Evans, 1988;

**Figure C.3.** Fat-free milk.

Darwish and DeLuca, 1996b). This is the age of biological knowledge, and molecular biology is having a profound impact on the study of physiological systems.

Calcium is vital to aging populations and to women's health in general (Heaney, 1996). Study of the regulation of calcium forces one to take into account nutrition, for the utilization of calcium occurs not in a vacuum but is linked to other minerals and proteins in the diet. The evolution of behavior and that of physiology converge in the regulation of calcium.

But there is a calcium fad out there. One has only to look at a recent advertising supplement in *Newsweek* to see how milk is being marketed, particularly fat-free milk (Figure C.3).

The important point is that calcium plays a fundamental role in the prevention of disease and the maintenance of end-organ systems. Nutritional needs change over the life cycle. With greater longevity, nutrition becomes more important. Calcium is an important mineral for general health and for sustaining the quality of life in the later years.

# Appendix

Let us look at calcium in food sources. This material is adapted from several sources, including the American Medical Association and others mentioned in the Introduction to this book. This list includes those discussed earlier, in addition to other sources of calcium. Again, consider the wide range of calcium contents in the foods that we ingest.

| Food | Serving size | Calcium (mg) |
|------|-------------|-------------|
| Milk | | |
| Whole, 4% | 1 cup | 300 |
| Nonfat | 1 cup | 300 |
| Buttermilk | 1 cup | 300 |
| Cheese | | |
| American | 1 oz. | 200 |
| Blue | 1 cu. in. | 50 |
| Cheddar | 1 oz. | 120 |
| Cottage, 2% fat | 1 cup | 200 |
| Parmesan, grated | 1/4 cup | 100 |
| Swiss | 1 oz. | 260 |
| Ice cream | | |
| Hard | 1 cup | 200 |
| Soft | 1 cup | 275 |
| Pudding | | |
| Chocolate | 1 cup | 250 |
| Vanilla | 1 cup | 300 |
| Custard | 1 cup | 300 |
| Yogurt | | |
| Whole | 1 cup | 300 |
| Low-fat-milk | 1 cup | 400 |
| Meat, poultry, seafood, beef | 2+ oz. | 10 |
| Chicken breast | 2+ oz. | 9 |
| Eggs | 1 | 25 |
| Oysters, raw | 1 cup | 225 |
| Salmon, pink, canned | 3 oz. | 150 |

(Cont.)

| Food | Serving size | Calcium (mg) |
|---|---|---|
| Sardines, canned in oil | 3 oz. | 375 |
| Shrimp, canned | 3 oz. | 100 |
| Tuna, canned in oil | 3 oz. | 7 |
| Beans (cooked) | | |
| Black beans | 1 cup | 150 |
| Garbanzo | 1 cup | 150 |
| Lima | 1 cup | 80 |
| Navy | 1 cup | 100 |
| Red kidney | 1 cup | 80 |
| Tofu (soybean curd) | 4 oz. | 150 |
| Whole sesame seeds | 3 tbsp. | 300 |
| Vegetables | | |
| Broccoli, cooked | 1 stalk | 150 |
| Brussels sprouts | 1 cup | 50 |
| Cabbage, cooked | 1 cup | 60 |
| Bok choy, cooked | 1 cup | 250 |
| Carrots | 1 cup | 50 |
| Cashew nuts | 1 cup | 50 |
| Cauliflower, cooked | 1 cup | 25 |
| Collards, cooked | 1 cup | 300 |
| Cucumber | 1 large | 75 |
| Mustard greens and kale | 1 cup | 200 |
| Okra, cooked | 1 cup | 150 |
| Onions, cooked | 1 cup | 50 |
| Parsley, raw | 1 cup | 120 |
| Peanuts, roasted | 1 cup | 100 |
| Sauerkraut, canned | 1 cup | 85 |
| Spinach | 1 cup | 200 |
| Squash, butternut | 1 cup | 80 |
| Turnip greens, cooked | 1 cup | 250 |
| Fruit | | |
| Apricots, dried, uncooked | 1 cup | 100 |
| Blackberries | 1 cup | 50 |
| Orange | 1 medium | 50 |
| Orange juice | 1 cup | 25 |
| Raisins | 1/3 cup | 30 |
| Rhubarb, cooked | 1 cup | 212 |
| Grains | | |
| Whole-wheat bread | 1 slice | 25 |
| Enriched farina | 1 cup | 150 |
| Pie, custard | 4-in. section | 125 |
| Pie, custard pumpkin | 4-in. section | 150 |
| Pizza, cheese | 5-in.+ section | 100 |

(Cont.)

| Food | Serving size | Calcium (mg) |
|------|-------------|-------------|
| Sweets | | |
|   Caramels | 1 oz. | 42 |
|   Molasses | 1 tbsp. | 150 |
|   Sugar, brown | 1 cup | 175 |
| Nuts | | |
|   Almonds | 1/2 cup | 150 |
|   Peanuts | 1 cup | 100 |
|   Walnuts | 1/2 cup | 150 |
| Mineral waters | | |
|   Evian | 1 liter | 78 |
|   Perrier | 1 liter | 140 |
|   Contrexeville | 1 liter | 451 |
|   San Pellegrino | 1 liter | 200 |
|   Ferrarelle | 1 liter | 446 |
|   Apolinaris | 1 liter | 91 |
|   Mountain Valley (U.S.) | 1 liter | 70 |
|   Mendocino (U.S.) | 1 liter | 380 |
| Foods highest in calcium | | |
|   Nonfat milk | 1 cup | 300 |
|   Baked custard | 1 cup | 300 |
|   Vanilla pudding | 1 cup | 300 |
|   Grilled cheese sandwich | 1 | 424 |
|   Ham/American cheese sandwich | 1 | 325 |
|   Cheddar, grated | 1/2 cup | 408 |
|   Parmesan cheese | 1 oz. | 336 |
|   Ricotta cheese, part skim | 1/2 cup | 337 |
|   Yogurt, fat-free, plain | 8 oz. | 452 |
|   Potato, baked with cheese & broccoli | 1 | 334 |
|   Sardines, with bones | Can | 375 |
|   Cooked collards | 1 cup | 300 |
|   Contrexeville water | 1 liter | 451 |
|   Ferrarelle water | 1 liter | 446 |
|   Mendocino water (U.S.) | 1 liter | 380 |
| Supplemental calcium | | |
|   Calcium carbonate | | 40% elemental calcium |
|   Calcium citrate | | 24% elemental calcium |
|   Calcium phosphate | | 23% elemental calcium |
|   Calcium lactate | | 13% elemental calcium |
|   Calcium gluconate | | 9% elemental calcium |

# Notes

## Chapter 1

1. To Bernard, however, that was not comprehensible. For Bernard, there was only one form of knowledge, and that was necessary knowledge. Bernard, like his predecessors, discounted statistics, and he offered this point: "If based on statistics, medicine can never be anything but a conjectural science; only by basing itself on experimental determinism can it become a true science" (1957, 139). Conjecture is at the heart of inquiry, as is statistical analysis, but during Bernard's time, conjecture was the less valued science. Over the past 150 years we have come to accept the legitimacy of probabilities in our scientific lexicon (Hacking, 1975). Today, most of us steeped in the methods of statistical design accept its prominence alongside conjecture, tested by biological and medical explanations. To rephrase Bernard for the modern world: If not based on rigorously collected empirical data analyzed by appropriate statistical methods, medicine can never be anything but an art, not a science.
2. Cannon was one of America's first world-class biological scientists, and in his laboratory of physiological sciences at Harvard University he covered much ground.
3. He is to psychobiology what Charles Peirce was to philosophy, what Benjamin Franklin and Thomas Jefferson were to early Americana – amazingly innovative, practical, and grounded.
4. Calcium sensors appear to be essential for plants to tolerate sodium. Excess sodium imposes a stress on plant growth. It can impair potassium regulation and inhibit proteins essential for plant life (Liu and Zhu, 1998). Calcium added to glycophytic plants will enhance sodium tolerance. A mutation in gene SOS3 creates vulnerability to sodium inhibition of plant growth (Liu and Zhu, 1998). What is important to note is that some plants, such as *Arabidopsis thaliana*, are loaded with both sodium and calcium and that calcium is essential for the plants to adapt to the high sodium content.

## Chapter 3

1. It is well known, for example, that in a variety of reptiles and birds, the hormones of calcium homeostasis are vital for mobilization of calcium-transport mechanisms and transport of calcium to and from the embryo and the eggshell (Packard and Clark, 1996). Like the fetus, the embryo lives in a rich calcium environment.

# References

Abrams, S.A., Stuff, J.E. (1994). Calcium metabolism in girls: current diets lead to low rates of calcium absorption and retention during pregnancy. *Am J Clin Nutr* 61:577–85.

Abrams, S.A., Wen, J., Stuff, J.E. (1997). Absorption of calcium, zinc and iron from breast milk by five-to-seven-month-old infants. *Pediatr Res* 41:384–90.

Abreo, K., Adlakha, A., Kilpatrick, S., Flanagan, R., Webb, R., Shakamuri, S. (1993). The milk-alkali syndrome. A reversible form of acute renal failure. *Arch Intern Med* 153:1005–10.

Adam, W.R. (1973). Novel diet preferences in potassium deficient rats. *J Comp Physiol Psychol* 84:286–8.

Adam, W.R., Dawborn, J.K. (1972). Effect of potassium depletion on mineral appetite in the rat. *J Comp Physiol Psychol* 78:51–8.

Adams, J.S., Lee, G. (1997). Gains in bone mineral density with resolution of vitamin D intoxication. *Ann Intern Med* 27:203–6.

Adams, P.H. (1988). Calcium-regulating hormones: general. Chapter 4 in *Calcium in Human Biology*, B.E.C. Nordin (ed.). New York: Springer-Verlag.

Affinito, P., Tommaselli, G.A., di Carlo, C., Guida, F., Nappi, C. (1996). Changes in bone mineral density and calcium metabolism in breastfeeding women. *J Clin Endocrinol Metab* 81:2314–18.

Ahearn, G.A., Zhuang, A. (1996). Cellular mechanisms of calcium transport in crustaceans. *Physiol Zool* 69:383–402.

Akar, N., Sipahi, T., Soyiemez, F. (1997). Coffee beans pica causing iron and zinc deficiency. *J Trace Elem Exp Med* 10:205–8.

Albright, F., Reifenstein, E.C. (1948). *Parathyroid Glands and Metabolic Bone Disease*. Baltimore: Williams & Wilkins.

Alderman, B.W., Weiss, N.S., Daling, J.R., Ure, C.L., Ballard, J.H. (1986). Reproductive history and postmenopausal risk of hip and forearm fracture. *Am J Epidemiol* 124:262–7.

Alfaham, M., Woodhead, S., Pask, G., Davies, D. (1995). Vitamin D deficiency: a concern in pregnant Asian women. *Br J Nutr* 73:881–7.

Alheid, G.F., Beltramino, C.A., Savory, J. (1993). A chronic increase in calcium but not sodium intake is observed after a single subcutaneous injection of lead chloride, but not by mercury chloride in female rats. *Soc Neurosci Abstr* 19:1676.

Alheid, G.F., de Olmos, J., Beltramino, C.A. (1996). Amygdala and extended amygdala. In: *The Rat Nervous System*, 2nd ed., G. Paxinos (ed.). San Diego: Academic Press.

Alheid, G.F., Schulkin, J., Epstein, A. (1992). Amygdala pathways involved in sodium appetite in the rat: anatomical considerations. *Soc Neurosci Abstr* 18:1231.

Allen, L.H. (1982). Calcium bioavailability and absorption: a review. *Am J Clin Nutr* 35:783–808.

Allen, M.E., Oftedal, O.T., Horst, R.I. (1995). Remarkable differences in the response to dietary vitamin D among species of reptiles and primates: Is ultraviolet B light essential? In: *Biological Effects of Light.* M.F. Holick, E.G. Jung (eds.). New York: W. de Gruyter.

Allen, M.E., Oftedal, O.T., Ullrey, D.E. (1993). Effect of dietary calcium concentration on mineral composition of fox geckos (*Hemidactylus garnoti*) and Cuban tree frogs (*Osteopilus septentrionalis*). *J Zoo Wild Med* 24:118–28.

Allen, S.H., Shah, J.H. (1992). Calcinosis and metastatic calcification due to vitamin D intoxication: a case report and review. *Horm Res* 37:68–77.

Aloia, J.F., Vaswani, A., Yeh, J.K., Ross, P.L., Flaster, E., Dilmanian, F.A. (1994). Calcium supplementation with and without hormone replacement therapy to prevent postmenopausal bone loss. *Ann Intern Med* 120:97–103.

American Medical Association. (1998). *Pocket Guide to Calcium.* New York: Random House.

Ammann, P., Irion, O., Gast, J., Bonjour, J.P., Beguin, F., Rizzoli, R. (1993). Alterations of calcium and phosphate metabolism in primary hyperparathyroidism during pregnancy. *Acta Obstet Gynecol Scand* 72:488–92.

Anderson, G.H. (1979). Control of protein and energy intake: role of plasma amino acid and brain neurotransmitters. *Can J Physiol Pharmacol* 57:1043–57.

Arai, R., Jacobowitz, D.M., Deura, S. (1994). Distribution of calretinin, calbinin-D28k, and parvalbumin in the rat thalamus. *Brain Res* 33:595–614.

Arai, R., Winsky, L., Arai, M.J., Jacobowitz, D.M. (1991). Immunohistochemical localization of calretinin in the rat hindbrain. *J Comp Neurol* 310:21–44.

Arnold, A.P., Breedlove, S.M. (1985). Organizational and activational effects of sex steroids on brain and behavior: a reanalysis. *Horm Behav* 19:469–98.

Atkinson, P.J., West, R.R. (1970). Loss of skeletal calcium in lactating women. *J Obstet Gynaecol Br Commonw* 77:555–60.

Atoni, F.A. (1997). Calcium regulation of adenylyl cyclase: relevance for endocrine control. *Trends in Endocrinology and Metabolism* 8:7–14.

Auerbach, G.D. (1988). Calcium-regulating hormones: parathyroid hormone and alcitonin. Chapter 3 in *Calcium in Human Biology*, B.E.C. Nordin (ed.). New York: Springer-Verlag.

Bachmanov, A.A., Tordoff, M.G., Beauchamp, G.K. (1998). Voluntary sodium chloride consumption by mice: differences among five inbred strains. *Behav Genet* 28:117–24.

Baeksgaard, L., Andersen, K.P., Hyldstrup, L. (1998). Calcium and vitamin D supplementation increases spinal BMD in healthy, postmenopausal women. *Osteoporos Int* 8:255–60.

Baer, D.J., Oftedal, O.T., Rumpler, W.V., Ullrey, D.E. (1997). Dietary fiber influences nutrient utilization, growth and dry matter intake of green iguanas (*Iguana iguana*). *J Nutr* 127:1501–7.

Baimbridge, K.G., Taylor, T.G. (1980). Role of calcitonin in calcium homeostasis in the chick embryo. *J Endocrinol* 85:171–85.

Baker, P.P., Mazess, R.B. (1963). Calcium: unusual sources in the Highland Peruvian diet. *Science* 142:1466–7.

Balabanova, S., Peter, J., Reinhardt, G. (1986). Parathyroid hormone secretion by brain and pituitary of sheep. *Klin Wochenschr* 64:173–6.

Baldwin, G.F., Bentley, P.J. (1980). Calcium metabolism in bullfrog tadpoles (*Rana cates-beiana*). *J Exp Biol* 88:357–65.

Baldwin, G.F., Bentley, P.J. (1981). A role for skin in calcium metabolism of frogs (*Rana pipiens*). *Comp Biochem Physiol A Comp Physiol* 68:181–6.

Bancroft, J., Cook, A., Williamson, L. (1988). Food craving, mood and the menstrual cycle. *Psychol Med* 18:855–60.

Barclay, R.M.R. (1994). Constraints on reproduction by flying vertebrates: energy and calcium. *Amer Nat* 144:1021–31.

Barclay, R.M.R. (1995). Does energy or calcium availability constrain reproduction by bats? *Symp Zool Soc London* 67:245–58.

Barger-Lux, M.J., Heaney, R.P. (1992). Nutritional correlates of low calcium intake. *Clin Appl Nutr* 2:39–44.

Barger-Lux, M.J., Heaney, RP. (1994). The role of calcium intake in preventing bone fragility, hypertension, and certain cancers. *J Nutr* 124(8S):1406S–11S.

Barger-Lux, M.J., Heaney, R.P. (1995). Caffeine and the calcium economy revisited. *Osteoporos Int* 5:97–102.

Barger-Lux, M.J., Heaney, R.P., Lanspa, S.J., Healy, J.C., DeLuca, H.F. (1995). An investigation of sources of variation in calcium absorption efficiency. *J Clin Endocrinol Metab* 80:406–11.

Baron, J. (1994). *Thinking and Deciding*, 2nd ed. Cambridge University Press. (Originally published 1988.)

Baron, J., Holzman, G.B., Schulkin, J. (1998). Attitudes of obstetricians and gynecologists toward hormone replacement therapy. *Medical Decision Making* 18:406–11.

Barr, S.I. (1994). Associations of social and demographic variables with calcium intakes of high school students. *J Am Diet Assoc* 94:260–79.

Barraclough, C.A. (1962). Studies on mating behaviour in the androgen-sterilized female rat in relation to the hypothalamic regulation of sexual behaviour. *J Endocrinol* 25:175–82.

Barrett, P.Q., Gertner, J.M., Rasmussen, H. (1980). Effect of dietary phosphate on transport properties of pig renal microvillus vesicles. *Am J Physiol* 239:F352–9.

Bartlet, J.P. (1985). Calcitonin may modulate placental transfer of calcium in ewes. *J Endocrinol* 104:17–22.

Bartness, T.J., Waldbillig, R.J. (1984). Dietary self-selection in intact, ovariectomized, and estradiol-treated female rats. *Behav Neurosci* 98:125–37.

Barton, J.R., Riely, C.A., Sibai, B.M. (1992). Baking powder pica mimicking preeclampsia. *Am J Obstet Gynecol* 167:98–9.

Bartoshuk, L.M. (1974). NaCl thresholds in man: thresholds for water taste or NaCl taste? *J Comp Physiol Psychol* 87:310–25.

Bartoshuk, L.M. (1988). Clinical psychophysics of taste. *Gerodontics* 4:249–55.

Bartoshuk, L.M. (1991). Taste, smell and pleasure. In: *The Hedonics of Taste*, R.C. Bolles (ed.). Hillsdale, NJ: Erlbaum.

Beatty, W.W., Powley, T.L., Keesey, R.E. (1970). Effects of neonatal testosterone injection and hormone replacement in adulthood on body weight and body fat in female rats. *Physiol Behav* 5:1093–8.

Beauchamp, G.K. (1987). The human preference for excess salt. *Am Scientist* 75:27–33.

Beauchamp, G.K., Bertino, M., Burke, D., Engelman, K. (1990). Experimental sodium depletion and salt taste in normal human volunteers, 1–3. *Am J Clin Nutr* 51:881–9.

**147**

Beauchamp, G.K., Bertino, M., Engelman, K. (1983). Modification of salt taste. *Ann Intern Med* 98:763–9.

Beck, M., Galef, B.G., Jr. (1989). Social influences on the selection of a protein-sufficient diet by Norway rats (*Rattus norvegicus*). *J Comp Psychol* 103:132–9.

Beebe-Center, J.G. (1932). *The Psychology of Pleasantness and Unpleasantness.* New York: Van Nostrand.

Behar, V., Nakamoto, C., Greenberg, Z., Bisello, A., Suva, L.J., Rosenblatt, M., Chorev, M. (1996). Histidine at position 5 is the specificity "switch" between two parathyroid hormone receptor subtypes. *Endocrinology* 137:4217–24.

Beidler, L.M. (1953). Properties of chemoreceptors of tongue of rat. *J Neurophysiol* 16:595–607.

Beidler, L.M., Fishman, I.Y., Hardiman, C.W. (1955). Species differences in taste responses. *J Neurophysiol* 181:235–9.

Belizan, J.M., Villar, J., Bergel, E., del Pino, A., DiFulvio, S., Galliano, S.V., Kattan, C. (1997). Long term effect of calcium supplementation during pregnancy on the blood pressure of offspring: follow up of a randomised controlled trial. *Br Med J* 313:281–5.

Belizan, J.M., Villar, J., Gonzalez, L., Campodonico, L., Bergel, E. (1991). Calcium supplementation to prevent hypertensive disorders of pregnancy. *N Engl J Med* 325:1399–405.

Belovsky, G.E. (1978). Diet optimization in a generalist herbivore: the moose. *Theor Popul Biol* 14:105–34.

Belovsky, G.E. (1986). Optimal foraging and community structure: implications for a guild of generalist grassland herbivores. *Oecologia* 70:35–52.

Bentley, P.J., Baldwin, G.F. (1980). Comparison of transcutaneous permeability in skins of larval and adult salamanders (*Ambystoma tigrinum*). *Am J Physiol* 239:R505–8.

Bernard, C. (1957). *An Introduction to the Study of Experimental Medicine*, trans. H.C. Greene. New York: Dover. (Originally published 1865.)

Bernard, R.T.F., Davison, A. (1996). Does calcium constrain reproductive activity in insectivorous bats? Some empirical evidence for Schreibers' long-fingered bat (*Miniopterus schreibersii*). *S Afr J Zool* 31(4):218–20.

Bernstein, I.L., Hennessy, C.J. (1987). Amiloride-sensitive sodium channels and expression of sodium appetite in rats. *Am J Physiol* 253:R371–4.

Berridge, K.C. (1996). Food reward: brain substrates of wanting and liking. *Neurosci Biobehav Rev* 20:1–25.

Berridge, K.C. (2000). Measuring hedonic impact in human infants and animals: microstructure of affective taste reactivity patterns. *Neurosci Biobehav Rev* 24:173–98.

Berridge, K.C., Flynn, F.W., Schulkin, J., Grill, H.J. (1984). Sodium depletion enhances salt palatability in rats. *Behav Neurosci* 98:652–60.

Berridge, K.C., Grill, H.J. (1983). Alternating ingestive and aversive consummatory responses suggest a two-dimensional analysis of palatability in rats. *Behav Neurosci* 97:563–73.

Berridge, K.C., Grill, H.J. (1984). Isohedonic tastes support a two-dimensional hypothesis of palatability. *Appetite* 5:221–31.

Berridge, K.C., Schulkin, J. (1989). Palatability shift of a salt-associated incentive during sodium depletion. *Q J Exp Psychol* 41:121–38.

Bertelloni, S., Baroncelli, G.I., Pelletti, A., Battini, R., Saggese, G. (1994). Parathyroid hormone-related protein in healthy pregnant women. *Calcif Tissue Int* 54:195–197.

Bidmon, H.J., Mayerhofer, A., Heiss, C., Bartke, A., Stumpf, W.E. (1991). Vitamin D

(Soltriol) receptors in the choroid plexus and ependyma: their species-specific presence. *Molec Cell Neurosci* 2:145–56.

Bidmon, H.J., Stumpf, W.E. (1994). Distribution of target cells for 1,25-dihydroxyvitamin D in the brain of the yellow-bellied turtle *Trachemys scripta*. *Brain Res* 640:277–85.

Bidmon, H.J., Stumpf, W.E. (1996). Vitamin D target systems in the brain of the green lizard *Anolis carolinensis*. *Anat Embryol* 193:145–60.

Bijvelds, M.J.C., van der Heijden, A.J.H., Flik, G., Verbost, P.M., Kolar, Z.I. Wendelaar-Bonga, S.E. (1997). Calcium pump activities in the kidneys of *Oreochromis mossambicus*. *J Exp Biol* 198:1351–7.

Bilezikian, J.P. (1992). Management of acute hypercalcemia. *N Engl J Med* 326:1196–203.

Bilezikian, J.P. (1993). Clinical reviews: management of hypercalcemia. *J Clin Endocrinol Metab* 77:1445–49.

Binderup, L. (1993). Comparison of calcipotriol with selected metabolites and analogues of vitamin $D_3$: effects on cell growth regulation in vitro and calcium metabolism in vivo. *Pharmacol Toxicol* 72:240–4.

Bird, E., Contreras, R.J. (1986). Maternal dietary sodium chloride levels affect the sex ratio in rat litters. *Physiol Behav* 36:307–10.

Bjornsson, B.T., Nilsson, S. (1985). Renal and extra renal excretion of calcium in the marine teleost, *Gadus morhua*. *Am J Physiol* 248:R18–22.

Black, A.E., Goldberg, G.R., Jebb, S.A., Livingstone, M.B., Cole, T.J., Prentice, A.M. (1991). Critical evaluation of energy intake data using fundamental principles of energy physiology: 2. Evaluating the results of published surveys. *Eur J Clin Nutr* 45:583–99.

Black, R.M., Weingarten, H.P., Epstein, A.N., Maki, R., Schulkin, J. (1992). Transection of the stria terminalis without damage to the medial amygdala does not alter behavioural sodium regulation. *Acta Neurobiol Exp* 52:9–15.

Blaine, E.H., Covelli, M.D., Denton, D.A., Nelson, J.F., Shulkes, A.A. (1975). The role of ACTH and adrenal glucocorticoids in the salt appetite of wild rabbits (*Oryctolagus cuniculus* (L)). *Endocrinology* 97:793–801.

Blair-West, J.R., Coghlan, J.P., Denton, D.A., Nelson, J.F., Orchard, E., Scoggins, B.A., Wright, R.D., Myers, K., Junqueira, C.L. (1968). Physiological, morphological and behavioural adaptation to a sodium deficient environment by wild native Australian and introduced species of animals. *Nature* 217:922–7.

Blair-West, J.R., Denton, D.A., McBurnie, M.I., Tarjan, E., Weisinger, R.S. (1995). Influence of adrenal steroid hormones on sodium appetite of BALB/c mice. *Appetite* 24:11–24.

Blair-West, J.R., Denton, D.A., McBurnie, M.I., Weisinger, R.S. (1996). The effect of adrenocorticotropic hormone on water intake in mice. *Physiol Behav* 60:1053–6.

Blair-West, J.R., Denton, D.A., McKinley, M.J., Radden, B.G., Ramshaw, E.H., Wark, J.D. (1992). Behavioral and tissue responses to severe phosphorus depletion in cattle. *Am J Physiol* 263:R656–63.

Blake, W.D., Jurf, A.N. (1968). Increased voluntary Na intake in K-deprived rats. *Communications in Behavioral Biology, A* 1:1–7.

Blundell, J.E. (1984). Systems and interactions: an approach to the pharmacology of eating and hunger. In: *Eating and Its Disorders*, A.J. Stunkard and E. Stellar (eds.), pp. 39–65. New York: Raven.

Bogan, A.D., DeWare, K.B. (1992). Calcium intake and knowledge of osteoporosis in university women. *Canadian Home Econ J* 42:80–4.

**149**

Boggess, K.A., Samuel, L., Schmucker, B.C., Waters, J., Easterling, T.R. (1997). A randomized controlled trial of the effect of third-trimester calcium supplementation on maternal hemodynamic function. *Obstet Gynecol* 90:157–67.

Bohlen, J.G. (1979). Biological rhythms: human responses in the polar environment. *Yearbook of Physical Anthropology* 22:47–79.

Bolles, R.C., Hayward, L., Crandall, C. (1981). Conditioned taste preferences based on caloric density. *J Exp Psychol Anim Behav Proc* 7:59–69.

Bonjour, J.P., Carrie, A.L., Ferrari, H., Clavien, G., Slosman, D., Theintz, G., Rizzoli, R. (1997). Calcium-enriched foods and bone mass growth in prepubertal girls: a randomized double-blind placebo-controlled trial. *J Clin Invest* 88:1297–5.

Booth, D.A. (1972). Conditioned satiety in the rat. *J Comp Physiol Psychol* 81:457–71.

Booth, D.A. (1974). Dietary flavor acceptance in infant rats established by association with effects of nutrient composition. *Physiol Psychol* 2:313–19.

Booth, D.A. (1985). Food-conditioned eating preferences and aversions with interoceptive elements: conditioned appetites and satieties. *Ann NY Acad Sci* 443:22–41.

Booth, D.A., and Simpson, P. (1971). Food preferences acquired by association with variation in amino acid nutrition. *Q J Exp Psychol* 23:135–45.

Bortell, R., Owen, T.A., Bidwell, J.P., Gavazzo, P., Breen, E., Van Wijnen, A.J., DeLuca, H.F., Stein, J.L., Lian, J.B., Stein, G.S. (1992). Vitamin D-responsive protein-DNA interactions at multiple promoter regulatory elements that contribute to the level of rat osteocalcin gene expression. *Proc Natl Acad Sci USA* 89:6119–23.

Bostick, R.M. (1997). Human studies of calcium supplementation and colorectal epithelial cell proliferation. *Cancer Epidemiol Biomarkers Prev* 6:971–80.

Botkin, D.B., Jordan, P.A., Dominski, A.S., Lowendorf, H.S., Hutchinson, G.E. (1973). Sodium dynamics in a northern ecosystem. *Proc Natl Acad Sci USA* 70:2745–8.

Bouchard, P., Roman, F., Junien, J.L., Quirion, R. (1996). Autoradiographic evidence for the modulation of in vivo sigma receptor labeling by neuropeptide Y and calcitonin related peptide in the mouse brain. *J Pharmacol Exp Ther* 276:223–30.

Bowden, S.J., Emly, J.F., Hughes, S.V., Powell, G., Ahmed, A., Whittle, M.J., Ratcliffe, J.G., Ratcliffe, W.A. (1994). Parathyroid hormone-related protein in human term placenta and membranes. *J Endocrinol* 142:217–24.

Bowen, D.J. (1992). Taste and food preference changes across the course of pregnancy. *Appetite* 19:233–42.

Bowen, D.J., Grunberg, N.E. (1990). Variations in food preference and consumption across the menstrual cycle. *Physiol Behav* 47:287–91.

Bradshaw, J.W.S., Neville, F., Sawyer, D. (1997). Factors affecting pica in the domestic cat. *Appl Anim Behav Sci* 52:373–379.

Braithwaite, G.D. (1978). The effect of dietary calcium intake of ewes in pregnancy on their calcium and phosphorus metabolism in lactation. *Br J Nutr* 39:213–18.

Brandwein, S.L., Sigman, K.M. (1994). Case report: milk-alkali syndrome and pancreatitis. *Am J Med Sci* 308:173–6.

Breedlove, S.M. (1992). Sexual dimorphism in the vertebrate nervous system. *J Neurosci* 12:4133–42.

Bregar, R.E., Strombakis, N., Allen, R.W., Schulkin, J. (1983). Brief exposure to a saline stimulus promotes latent learning in the salt hunger system. *Soc Neurosc Abstr*.

Brenza, H.L., Kimmel-Jehan, C., Jehan, F., Shinki, T., Wakino, S., Anazawa, H., Suda, T., DeLuca, H.F. (1998). Parathyroid hormone activation of the 25-hydroxyvitamin $D_3$-1-alphahydroxylase gene promoter. *Proc Natl Acad Sci USA* 95:1387–91.

**150**

Breslau, N.A. (1996a). Calcium, magnesium, and phosphorus: intestinal absorption. Chapter 7 in *Primer on the Metabolic Bone Diseases and Disorders of Mineral Metabolism*, 3rd ed., M.J. Favus (ed.). New York: Lippincott-Raven.

Breslau, N.A. (1996b). Calcium, magnesium, and phosphorus: renal handling and urinary excretion. Chapter 8 in *Primer on the Metabolic Bone Diseases and Disorders of Mineral Metabolism*, 3rd ed., M.J. Favus (ed.). New York: Lippincott-Raven.

Breslin, P.A.S., Beauchamp, G.K. (1995). Suppression of bitterness by sodium: variation among bitter taste stimuli. *Chem Senses* 20:609–23.

Breslin, P.A.S., Spector, A.C., Grill, H.J. (1993). Chorda tympani section decreases the cation specificity of depletion-induced sodium appetite in rats. *Am J Physiol* 264:R319–23.

Bridges, R.S., Numan, M., Ronsheim, P.M., Mann, P.E., Lupini, C.E. (1990). Central prolactin infusions stimulate maternal behavior in steroid-treated, nulliparous female rats. *Proc Natl Acad Sci USA* 87:8003–7.

Bro, S., Olgaard, K. (1997). Effects of excess parathyroid hormone on nonclassical target organs. *Am J Kidney Dis* 30:606–20.

Brommage, R., DeLuca, H.F. (1984). Self-selection of a high calcium diet by vitamin D-deficient lactating rats increases food consumption and milk production. *J Nutr* 114:1377–85.

Brommage, R., Jarnagin, K., DeLuca, H.F. (1984). 1,25-Dihydroxyvitamin $D_3$ normalizes maternal food consumption and pup growth in rats. *Am J Physiol* 246:E227–31.

Bronner, F. (1991). Calcium transport across epithelia. *Int Rev Cytol* 131:169–212.

Bronner, F. (1998). Calcium absorption – a paradigm for mineral absorption. *J Nutr* 128:917–20.

Brown, E.M., Chattopadhyay, N., Vassilev, P.M., Hebert, S.C. (1998). The calcium-sensing receptor (CaR) permits $Ca^{2+}$ to function as a versatile extracellular first messenger. *Recent Prog Horm Res* 53:257–80.

Brown, J.E., Toma, R.B. (1986). Taste changes during pregnancy. *Am J Clin Nutr* 43:414–18.

Brun, P., Dupret, J.M., Perret, C., Thomasset, M., Mathieu, H. (1987). Vitamin D-dependent calcium-binding proteins (CaBPs) in human fetuses: comparative distribution of 9K CaBP mRNA and 28K CaBP during development. *Pediatr Res* 21:362–7.

Buffenstein, R., Poppitt, S.D., McDevitt, R.M., Prentice, A.M. (1977). Food intake and the menstrual cycle: a retrospective analysis, with implications for appetite research. *Physiol Behav* 56:1067–77.

Bukoski, R.D., Kremer, D. (1991). Calcium-regulating hormones in hypertension: vascular actions. *Am J Clin Nutr* 54:220S–6S.

Bullamore, J.R., Wilkinson, R., Gallagher, J.C., Nordin, B.E., Marshall, D.H. (1970). Effect of age on calcium absorption. *Lancet* 2:535–7.

Buntin, J.D. (1992). Neural substrates for prolactin-induced changes in behavior and neuroendocrine function. *Poult Sci Reg* 4:275–87.

Burmester, J.K., Maeda, N., DeLuca, H.F. (1988). Isolation and expression of rat 1,25-dihydroxyvitamin $D_3$ receptor cDNA. *Proc Natl Acad Sci USA* 85:1005–9.

Bursey, R.G., Watson, M.L. (1983). The effect of sodium restriction during gestation on offspring brain development in rats. *Am J Clin Nutr* 37:43–51.

Burton, P.B., Moniz, C., Quirke, P., Malik, A., Bui, T.D., Juppuer, H., Segre, G.U., Knight, D.E. (1992). Parathyroid hormone-related peptide expression in fetal and neonatal development. *J Pathol* 167:291–6.

Bushinsky, D.A. (1998). Homeostasis and disorders of calcium and phosphorus concentration. In: *Primer on Kidney Diseases*, A. Greenberg (ed.), pp. 106–13. San Diego: Academic.

Bushinsky, D.A., Monk, R.D. (1998). Calcium. *Lancet* 352:306–11.

Butera, P.C., Xiong, M., Davis, R.J., Platania, S.P. (1996). Central implants of dilute estradiol enhance the satiety effect of CCK-8. *Behav Neurosci* 110:823–30.

Byrne, J., Thomas, M.R., Chan, G.M. (1987). Calcium intake and bone density of lactating women in their late childbearing years. *J Am Diet Assoc* 87:883–7.

Cabanac, M. (1971). Physiological role of pleasure. *Science* 17:1103–7.

Cabanac, M. (1979). Sensory pleasure. *Q Rev Biol* 34:1–29.

Callinan, V., O'Hare, J.A. (1998). Cardboard chewing: cause and effect of iron-deficiency anemia. *Am J Med* 85:449.

Calvo, M.S., Eastell, R., Offord, K.P., Bergstrahl, E.J., Burritt, M.F. (1991). Circadian variation in ionized calcium and intact parathyroid hormone: evidence for sex differences in calcuim homeostasis. *J Clin Endocrinol Metab* 72:69–76.

Calvo, M.S., Heath, H. (1988). Acute effects of oral phosphate-salt ingestion on serum phosphorus, serum ionized calcium, and parathyroid hormone in young adults. *Am J Clin Nutr* 47:1025–9.

Campbell, A.K. (1988). Calcium as an intracellular regulator. Chapter 11 in *Calcium in Human Biology*, B.E.C. Nordin (ed.). New York: Springer-Verlag.

Campos, R.V., Asa, S.L., Drucker, D.J. (1991). Immunocytochemical localization of parathyroid hormone-like peptide in rat fetus. *Cancer Res* 51:6351–7.

Cannon, W.B. (1929). *Bodily Changes in Pain, Hunger, Fear and Rage*, 2nd ed. New York: Harper & Row. (Originally published 1915.)

Cannon, W.B. (1966). *The Wisdom of the Body*, 2nd ed. New York: Norton. (Originally published 1932.)

Cappuccio, F.P., Elliott, P., Allender, P.S., Pryer, J., Follman, P.A., Cutler, J.A. (1995). Epidemiologic association between dietary calcium intake and blood pressure: a meta-analysis of published data. *Am J Epidemiol* 142:935–45.

Carafoli, E. (1987). Intracellular calcium homeostasis. *Annu Rev Biochem* 56:395–433.

Care, A.D. (1989). Development of endocrine pathways in the regulation of calcium homeostasis. *Baillieres Clin Endocrinol Metab* 3:671–88.

Care, A.D., Caple, I.W., Abbas, S.K., Pickard, D.W. (1986). The roles of the parathyroid and thyroid glands on calcium homeostasis in the ovine fetus. In: *The Physiological Development of the Fetus and Newborn*, C.T. Jones, P.W. Nahanielsz (eds.), pp. 135–40. London: Academic Press.

Carlberg, C., Bendik, I., Wyss, A., Meier, E., Sturzenbecker, L.J., Grippo, J.F., Hunziker, W. (1993). Two nuclear signalling pathways for vitamin D. *Nature* 361:657–60.

Carpenter, D.O., Briggs, D.B., Knox, A.P., Strominger, N. (1984). Peptide-induced emesis in dogs. *Behav Brain Res* 11:277–81.

Carpenter, D.O., Briggs, D.B., Knox, A.P., Strominger, N. (1988). Excitation of area postrema neurons by transmitters, peptides, and cyclic nucleotides. *J Neurophysiol* 59:358–69.

Carpenter, D.O., Briggs, D.B., Strominger, N. (1983). Responses of neurons of canine area postrema to neurotransmitters and peptides. *Cell Mol Neurobiol* 3:113–26.

Carpenter, S.J. (1982). Enhanced teratogenicity of orally administered lead in hamsters fed diets deficient in calcium or iron. *Toxicology* 24:259–72.

Carroll, K.K., Jacobson, E.A., Eckel, L.A., Newmark, H.L. (1991). Calcium and carcinogenesis of the mammary gland. *Am J Clin Nutr* 54:2065–85.

Carruth, B.R., Skinner, J.D. (1991). Practitioners beware: regional differences in beliefs about nutrition during pregnancy. *J Am Diet Assoc* 91:435–40.

Catalanotto, F.A. (1978). Effects of dietary methionine supplementation on preferences for NaCl solutions. *Behav Biol* 24:457–66.

Chait, A., Suaudeau, C., DeBeaurepaire, R. (1995). Extensive brain mapping of calcitonin-induced anorexia. *Brain Res Bull* 36:467–72.

Chamberlin, N.L., Saper, C.B. (1994). Topographic organization of respiratory responses to glutamate miscrostimulation of the parabrachial nucleus in the rat. *J Neurosci* 14:6500–10.

Chan, G.M., McMurry, M., Westover, K., Engelbert-Fenton, K., Thomas, M.R. (1982a). Effects of increased dietary calcium intake upon the calcium and bone mineral status of lactating adolescent and adult women. *Am J Clin Nutr* 46:319–23.

Chan, G.M., Ronald, N., Slater, P., Hollis, J., Thomas, R.M. (1982b). Decreased bone mineral status in lactating adolescent mothers. *J Pediatr* 101:767–70.

Chan, K.K., Robinson, G., Pipkin, F.B. (1997). Differential sensitivity of human non-pregnant and pregnant myometrium to calcitonin gene-related peptide. *J Soc Gynecol Invest* 4:15–21.

Chapuy, M.C., Arlot, M.E., Duboeut, F., Brun, J., Crouzet, B., Arnaud, S., Delmas, P.D., Meunier, P.J. (1992). Vitamin $D_3$ and calcium to prevent hip fractures in elderly women. *N Engl J Med* 327:1637–42.

Chapuy, M.C., Chapuy, P., Thomas, J.L., Hazard, M.C., Meunier, P.J. (1996). Biochemical effects of calcium and vitamin D supplementation in elderly, institutionalized, vitamin-D-deficient patients. *Rev Rhum Engl Ed* 63:135–40.

Charles, P., Eriksen, E.F., Hasling, C., Sondergard, K., Mosekilde, L. (1991). Dermal, intestinal, and renal obligatory losses of calcium: relation to skeletal calcium loss. *Am J Clin Nutr* 54:266S–73S.

Chattopadhyay, N., Legradi, G., Bai, M., Kifor, O., Ye, C., Vassilev, P.M., Brown, E.M., Lechan, R.M. (1997a). Calcium-sensing receptor in the rat hippocampus: a developmental study. *Dev Brain Res* 100:13–21.

Chattopadhyay, N., Vassilev, P.M., Brown, E.M. (1997b). Calcium-sensing receptor: roles in and beyond systemic calcium homeostasis. *Biol Chem* 378:759–68.

Chawla, S., Hardingham, G.E., Quinn, D.R., Bading, H. (1998). CBP: a signal-regulated transcriptional coactivator controlled by nuclear calcium and CaM kinase IV. *Science* 281:1505–9.

Chel, V.G., Ooms, M.E., Popp-Snijders, C., Pavel, S., Schothorst, A.A., Meulemans, C.C., Lips, P. (1998). Ultraviolet irradiation corrects vitamin D deficiency and suppresses secondary hyperparathyroidism in the elderly. *J Bone Miner Res* 13:1238–42.

Chen, S.W., Francis, B.M., Dziuk, P.J. (1993). Effect of concentration of mixed-function oxidase on concentration of estrogen, rate of egg lay, eggshell thickness, and plasma calcium in laying hens. *J Anim Sci* 71:2700–7.

Chesters, J.K., Quarterman, J. (1970). Effects of zinc deficiency on food intake and feeding patterns of rats. *Br J Nutr* 24:1061.

Chiaraviglio, E. (1971). Amygdaloid modulation of sodium chloride and water intake in the rat. *J Comp Physiol Psychol* 76:401–7.

Chomsky, N. (1972). *Language and Mind.* New York: Harcourt Brace Jovanovich.

Chow, S.Y., Sakai, R.R., Witcher, J.A., Adler, N.T., Epstein, A. (1992). Sex and sodium intake in the rat. *Behav Neurosci* 106:172–80.

Christakos, S., Malkowitz, L., Sori, A., Sperduto, A., Feldman, S.C. (1987). Calcium binding protein in squid brain: biochemical similarity to the 28,000-M vitamin D-dependent calcium binding protein (calbindin-D 28K). *J Neurochem* 49:1427–37.

Christensen, C.M., Caldwell, D.F., Oberleas, D. (1974). Establishment of a learned preference for a zinc-containing solution by zinc-deficient rats. *J Comp Physiol Psychol* 87:415–21.

Christopoulos, G., Paxinos, G., Huang, X.F., Beaumont, K., Toga, A.W., Sexton, P.M. (1995). Comparative distribution of receptors for amylin and the related peptides calcitonin gene related peptide and alcitonin in rat and monkey brain. *Can J Physiol Pharmacol* 73:1037–41.

Clark, S.A., Boass, A., Toverud, S.U. (1987). Effects of high dietary contents of calcium and phosphorus on mineral metabolism and growth of vitamin D-deficient suckling and weaned rats. *Bone Miner* 2:257–70.

Clarke, S.N., Ossenkopp, K.P. (1998). Taste reactivity responses in rats: influence of sex and the estrous cycle. *Am J Physiol* 274:R718–24.

Clarkson, P.M., Haymes, E.M. (1995). Exercise and mineral status of athletes: calcium, magnesium, phosphorus, and iron. *Med Sci Sports Exerc* 27:831–43.

Clements, T.L., Zhou, X.Y., Myles, M., Endres, D., Lindsay, R. (1986). Serum vitamin $D_2$ and vitamin $D_3$ concentrations and absorption of vitamin $D_2$ in elderly subjects. *J Clin Endocrinol Metab* 63:656–60.

Coirini, H., Schulkin, J., McEwen, B.S. (1988). Behavioural and neuroendocrine regulation of mineralocorticoid and glucocorticoid actions. *Soc Neurosci Abstr*.

Coldwell, S.E., Tordoff, M.G. (1993). Latent learning about calcium and sodium. *Am J Physiol* 265:R1480–4.

Coldwell, S.E., Tordoff, M.G. (1996a). Acceptance of minerals and other compounds by calcium-deprived rats, 24-h tests. *Am J Physiol* 271:R1–10.

Coldwell, S.E., Tordoff, M.G. (1996b). Immediate acceptance of minerals and HCl by calcium-deprived rats: brief exposure tests. *Am J Physiol* 271:R11–17.

Cole, D.E.C., Peltekova, V.D., Rubin, L.A., Hawker, G.A., Vieth, R., Liew, C.C., Hwang, D.M., Evrovski, J., Hendy, G.N. (1999). A986S polymorphism of the calcium-sensing receptor and circulating calcium concentrations. *Lancet* 353:112.

Collier, G., Leshner, A.I., Squibb, R.L. (1969a). Dietary self-selection in active and non-active rats. *Physiol Behav* 4:79–82.

Collier, G., Leshner, A.I., Squibb, R.L. (1969b). Self-selection of natural and purified dietary protein. *Physiol Behav* 4:83–6.

Committee on Measuring Lead in Critical Populations (1993). *Measuring Lead Exposure in Infants, Children, and Other Sensitive Populations*. Washington, DC: National Academy Press.

Contreras, R.J. (1977). Changes in gustatory nerve discharges with sodium deficiency: a single unit analysis. *Brain Res* 121:373–8.

Contreras, R.J. (1989). Gustatory mechanisms of a specific appetite. In: *Neural Mechanisms in Taste*, R.H. Cagan (ed.), pp. 119–46. Boca Raton: CRC Press.

Cooke, H.M., Tarbnia, G., Breedlove, M. (1999). A brain sexual dimorphism controlled by adult circulating androgens. *Proc Natl Acad Sci USA* 96:7538–40.

Cooksey, N.R. (1995). Pica and olfactory craving of pregnancy: how deep are the secrets? *Birth* 22:129–37.

154

Coppock, C.E. (1970). Free choice mineral consumption by dairy cattle. In: *Proceedings of the Cornell Nutrition Conference for Feed Manufacturers*, pp. 29–35.

Coppock, C.E., Evertt, R.W., Belyea, R.L. (1976). Effect of low calcium or low phosphorus diets on free choice consumption of dicalcium phosphate by lactating dairy cows. *J Dairy Sci* 59:571–80.

Coronios-Vargas, M., Toma, R.B., Tuveson, R.V., Shutz, I.M. (1992). Cultural influences on food cravings and aversions during pregnancy. *Ecol Food Nutr* 27:43–9.

Cortes, R., Arvidsson, U., Schalling, M., Ceccatelli, S., Hokfelt, T. (1990). In situ hybridization studies on mRNAs for cholecystokinin, calcitonin gene-related peptide and choline acetyltransferase in the lower brain stem, spinal cord and dorsal root ganglia of rat and guinea pig with special reference to motoneurons. *J Chem Neuroanat* 3:467–85.

Cowan, I.M., Brink, V.C. (1949). Natural game licks in the Rocky Mountain national parks of Canada. *J Mammol* 30(4):379–87.

Craig, W.C. (1918). Appetites and aversions as consitutents of instincts. *Biol Bull* 34:96–103.

Crichton-Browne, J. (1910). *Delusions in Diet*. London: Funk & Wagnalls.

Crooks, A.J., Simkiss, K. (1975). Calcium transport by the chick chorioallantois in vivo. *Q J Exp Physiol* 60:55–63.

Cross, N.A., Hillman, L.S., Allen, S.H., Krause, G.F. (1995a). Changes in bone mineral density and markers of bone remodeling during lactation and postweaning in women consuming high amounts of calcium. *J Bone Miner Res* 10:1312–20.

Cross, N.A., Hillman, L.S., Allen, S.H., Krause, G.F., Vieira, N.E. (1995b). Calcium homeostasis and bone metabolism during pregnancy, lactation, and postweaning: a longitudinal study. *Am J Clin Nutr* 61:514–23.

Cruikshank, D.P., Pitkin, R.M., Reynolds, W.A., Williams, G.A., Hargis, G.K. (1980a). Altered maternal calcium homeostasis in diabetic pregnancy. *J Clin Endocrinol Metab* 50:264.

Cruikshank, D.P., Pitkin, R.M., Reynolds, W.A., Williams, G.A., Hargis, G.K. (1980b). Calcium-regulating hormones and ions in amniotic fluids. *Am J Obstet Gynecol* 136:621–5.

Crystal, S.R., Bernstein, I.L. (1995). Morning sickness: impact on offspring salt preference. *Appetite* 25:231–40.

Csaja, J.A. (1975). Food rejection by female rhesus monkeys during the menstrual cycle and early pregnancy. *Physiol Behav* 14:579–87.

Dacke, C.G. (1979). *Calcium Regulation in Sub-Mammalian Vertebrates*. New York: Academic Press.

Dahlman, T., Sjoberg, E.S., Bucht, E. (1994). Calcium homeostasis in normal pregnancy and puerperium. *Acta Obstet Gynecol Scand* 73:393–8.

Dallman, M.F., Akana, S.F., Scribner, K.A., Bradbury, M.J., Walker, C.D., Strack, A.M., Cascio, C.S. (1992). Stress, feedback and facilitation in the hypothalamic-pituitary-adrenal axis. *J Neuroendocrinol* 4:517–26.

Dallman, M.F., Akana, S.F., Strack, A.M., Hanson, E.S., Sebastian, R.J. (1995). The neural network that regulates energy balance is responsive to glucocorticoids and insulin and also regulates HPA axis responsivity at a site proximal to CRF neurons. *Ann NY Acad Sci* 771:730–42.

Danielson, J., Buggy, J. (1980). Depression of ad lib and angiotensin-induced sodium intake at oestrus. *Brain Res Bull* 5:501–4.

Danks, J.A., Trivett, M.K., Power, D.M., Canario, A.V., Martin, T.J., Ingleton, P.M. (1998). Parathyroid hormone-related protein in lower vertebrates. *Clin Exp Pharmacol Physiol* 25:750–2.

Darwin, C. (1901). *Descent of Man.* London: Rand McNally. (Originally published 1871.)

Darwin, C. (1958). *The Origin of Species.* New York: New American Library (Mentor Books). (Originally published 1859.)

Darwin, C. (1965). *The Expression of the Emotions in Man and other Animals.* University of Chicago Press. (Originally published 1872.)

Darwish, H.M., DeLuca, H.F. (1996a). Analysis of binding of the 1,25-dihydroxyvitamin $D_3$ receptor to positive and negative vitamin D response elements. *Arch Biochem Biophys* 334:223–34.

Darwish, H.M., DeLuca, H.F. (1996b). Recent advances in the molecular biology of vitamin D action. *Prog Nucl Acid Res Mol Biol* 53:321–44.

Davies, M., Adams, P.H. (1978). The continuing risk of vitamin D intoxication. *Lancet* 2:621–3.

Davis, C.M. (1928). Self-selection of diet by newly weaned infants; an experimental study. *Am J Dis Child* 36:651–79.

Davis, C.M. (1939). Results of the self-selection of diets by young children. *Can Med Assoc J* 41:257–61.

Davis, J.W., Ross, P.D., Johnson, N.E., Wasnich, R.D. (1995). Estrogen and calcium supplement use among Japanese-American women: effects upon bone loss when used singly and in combination. *Bone* 17:369–73.

Dawson-Hughes, B. (1991). Calcium supplementation and bone loss: a review of controlled clinical trials, 1–3. *Am J Clin Nutr* 54:274S–80S.

Dawson-Hughes, B., Dallal, G.E., Krall, E.A., Harris, S., Sokoll, L.J., Falconer, G. (1991). Effect of vitamin D supplementation on wintertime and overall bone loss in healthy menopausal women. *Ann Intern Med* 115:505–12.

Dawson-Hughes, B., Dallal, G.E., Krall, E.A., Sadowski, L., Sahyoun, N., Tannenbaum, S.A. (1990). A controlled trial of the effect of calcium supplementation on bone density in postmenopausal women. *N Engl J Med* 323:878–83.

Dawson-Hughes, B., Harris, S.S., Krall, E.A., Dallal, G. (1997). Effect of calcium and vitamin D supplementation on bone density in men and women 65 years of age or older. *N Engl J Med* 337:670–6.

Dawson-Hughes, B., Harris, S.S., Krall, E.A., Dallal, G.E., Falconer, G., Green, C.L. (1995). Rates of bone loss in postmenopausal women randomly assigned to one of two dosages of vitamin D. *Am J Clin Nutr* 61:1140–5.

Dawson-Hughes, B., Harris, S., Kramich, C., Dallal, G., Rasmussen, H.M. (1993). Calcium retention and hormone levels in black and white women on high- and low-calcium diets. *J Bone Miner Res* 8:779–87.

Dawson-Hughes, B., Stern, D.T., Shipp, C.C., Rasmussen, H.M. (1988). Effect of lowering dietary calcium intake on fractional whole body calcium retention. *J Clin Endocrinol Metab* 67:62–8.

Deftos, L.J. (1996). Calcitonin. Chapter 14 in *Primer on the Metabolic Bone Diseases and Disorders of Mineral Metabolism*, 3rd ed., M.J. Favus (ed.). New York: Lippincott-Raven.

Delay, R.J., Kinnamon, S.C., Roper, S.D. (1997). Serotonin modulates voltage-dependent calcium current in *Necturus* taste cells. *J Neurophysiol* 77:2515–24.

Del Canho, S., Eilon, R., Schulkin, J., Leshem, M. (1996). "Need-free" and "need-induced" calcium preference during the lifespan of rats. *Isr J Med Sci* 32:S29 (abstract).

DeLuca, H.F. (1985). The vitamin D–calcium axis – 1983. In: *Calcium in Biological Systems*, R.P. Rubin, G.B. Weiss, J.W. Putney (eds.). New York: Plenum.

DeLuca, H.F. (1988). The vitamin D story: a collaborative effort of basic science and clinical medicine. *FASEB J* 2:224–36.

DeLuca, H.F. (1992). New concepts of vitamin D functions. *Ann NY Acad Sci* 669:68–9.

DeLuca, H.F. (1995). Aging, the vitamin D system and osteoporosis. In: *Adaptations in Aging*. New York: Academic Press.

Delvin, E.E., Gilbert, M., Pere, M.C., Garel, J.M. (1988). In vivo metabolism of calcitriol in the pregnant rabbit doe. *J Dev Physiol* 10:451–9.

Denton, D.A. (1965). Evolutionary aspects of the emergence of aldosterone secretion and salt appetite. *Physiol Rev* 45:245–95.

Denton, D.A. (1982). *The Hunger for Salt*. New York: Springer-Verlag.

Denton, D.A., Blair-West, J.R., McKinley, M.J., Nelson, J.F. (1986). Physiological analysis of bone appetite (osteophagia). *Bioessays* 4:40–2.

Denton, D.A., Nelson, J.F. (1971). The effects of pregnancy and lactation on the mineral appetites of wild rabbits (*Oryctolagus cuniculus* L.). *Endocrinology* 88:31–40.

Denton, D.A., Sabine, J.R. (1961). The selective appetite for Na, shown by Na-deficient sheep. *J Physiol* 157:97–116.

Desor, J.A., Maller, O., Andrews, K. (1975). Ingestive responses of human newborns to salty, sour, and bitter stimuli. *J Comp Physiol Psychol* 89(8):966–70.

Dethier, V.G. (1968). Chemosensory input and taste discrimination in the blowfly. *Science* 161:389–91.

Dethier, V.G. (1976). *The Hungry Fly*. Cambridge, MA: Harvard University Press.

Dethier, V.G., Stellar, E. (1961). *Animal Behavior*. Englewood Cliffs, NJ: Prentice-Hall.

Deutsch, J.A., Moore, B.O., Heinrichs, S.C. (1989). Unlearned specific appetite for protein. *Physiol Behav* 46:619–24.

Devenport, L.D. (1973). Aversion to a palatable saline solution in rats: interactions of physiology and experience. *J Comp Physiol Psychol* 83:95–105.

Devine, A., Criddle, R.A., Dick, I.M., Kerr, D.A., Prince, R.L. (1995). A longitudinal study of the effect of sodium and calcium intakes on regional bone density in postmenopausal women. *Am J Clin Nutr* 62:740–5.

Devine, A., Prince, R.L., Kerr, D.A., Dick, I.M., Criddle, R.A., Kent, G.N., Prince, R.L., Webb, P.G. (1993). Correlates of intestinal calcium absorption in women 10 years past the menopause. *Calcif Tissue Int* 52:358–60.

Dial, J., Avery, D.D. (1991). The effects of pregnancy and lactation on dietary self-selection in the rat. *Physiol Behav* 49:811–13.

Diaz, R., Hurwitz, S., Chattopadhyay, N., Pines, M., Yank, Y., Kifor, O., Einat, M.S., Butters, R., Hebert, S.C., Brown, E.M. (1997). Cloning, expression, and tissue localization of the calcium-sensing receptor in chicken (*Gallus domesticus*). *Am J Physiol* 273:R1008–16.

Dickinson, A. (1986). Re-examination of the role of the instrumental contingency in the sodium appetite irrelevant incentive effect. *Q J Exp Psychol* 38:161–72.

DiLorenzo, P.M., Monroe, S. (1989). Taste responses in the parabrachial pons of male, female and pregnant rats. *Brain Res* 23:219–27.

Dippel, R.L., Elias, J.W. (1980). Preferences for sweets in relationship to use of oral contraceptives and pregnancy. *Horm Behav* 24:1–6.

Donangelo, C.M., Trugo, N.M.F., Melo, G., Gomes, D.D., Henriques, C. (1996). Calcium homeostasis during pregnancy and lactation in primiparous and multiparous women with sub-adequate calcium intakes. *Nutr Res* 16:1631–40.

Donelson, E., Nims, B., Hunscher, H.A., Macy, I.G. (1931). Metabolism of women during the reproductive cycle. IV. Calcuim and phosphorus utilization in late lactation and during subsequent reproductive rest. *J Biol Chem* 91:675–86.

Doris, P.A. (1988). Plasma angiotensin II: interdependence on sodium and calcium homeostasis. *Peptides* 9:243–8.

Doussau, F., Clabecq, A., Henry, J.P., Darchen, F., Poulain, B. (1998). Calcium-dependent regulation of rab3 in short-term plasticity. *J Neurosci* 18:3147–57.

Dove, W.F. (1935). A study of individuality in the nutritive instincts and of the causes and effects of variations in the selection of foods. *Amer Nat* 69:469–544.

Drinkwater, B.L., Bruemner, B., Chestnut, C. (1990). Menstrual history as a determinant of current bone density in young athletes. *JAMA* 263:545–8.

Drinkwater, B.L., Chestnut, C.H., III. (1991). Bone density changes during pregnancy and lactation in active women: a longitudinal study. *Bone Miner* 14:153–60.

Duffy, V.B., Bartoshuk, L.M., Striegel-Moore, R., Rodin, J. (1998). Taste changes across pregnancy. *Ann NY Acad Sci* 885:805–9.

Duncan, C.J. (1962). Salt preferences of birds and mammals. *Physiol Zool* 35:120–32.

Dunn, C., Kolasa, K., Dunn, P.C., Ogle, M.B. (1994). Dietary intake of pregnant adolescents in a rural southern community. *J Am Diet Assoc* 94:1040–1.

Durand, D., Bartlet, J.P., Braithwaite, G.D. (1983). The influence of 1,25-dihydroxy-cholecalciferol on the mineral content of foetal guinea-pigs. *Reprod Nutr Devel* 23:235–44.

Dusso, A.S., Brown, A.J. (1998). Mechanism of vitamin D action and its regulation. *Am J Kidney Dis* 32:S13–S24.

Dwyer, J.H., Dwyer, K.M., Scribner, R.A., Sun, P., Li, L., Nicholson, L.M., Davis, I.J., Hohn, A.R. (1998). Dietary calcium, calcium supplementation, and blood pressure in African American adolescents. *Am J Clin Nutr* 68:648–55.

Dye, L., Blundell, J.E. (1997). Menstrual cycle and appetite control: implications for weight regulation. *Hum Reprod* 12:1142–51.

Dye, L., Warner, P., Bancroft, J. (1995). Food craving during the menstrual cycle and its relationship to stress, happiness of relationship and depression; a preliminary enquiry. *J Affect Disord* 34:157–64.

Eaton, S.B., Konner, M. (1985). Paleolithic nutrition. A consideration of its nature and current implications. *N Engl J Med* 312:283–9.

Eaton, S.B., Nelson, D.A. (1991). Calcium in evolutionary perspective. *Am J Clin Nutr* 54:281S–7S.

Ebashi, S. (1988). Cellular calcium: muscle. Chapter 12 in *Calcium in Human Biology*, B.E.C. Nordin (ed.). New York: Springer-Verlag.

Ebeling, P.R., Yergey, A.L., Vieira, N.E., Burrit, M.F., O'Fallon, W.M., Kumar, R., Riggs, B.L. (1994). Influence of age on effects of endogenous 1,25-dihydroxyvitamin D on calcium absorption in normal women. *Calcif Tissue Int* 55:330–4.

Eck, L.H., Hackett-Renner, C. (1992). Calcium intake in youth: sex, age, and racial differences in NHANES 11. *Prev Med* 21:473–82.

Eckert, J.F. (1938). Further observations on the calcium appetite of parathyroidectomized rats. *Am J Physiol* 123:59.

Edgren, R.A. (1960). A seasonal change in bone density in female musk turtles, *Sternothaerus odoratus* (Latreille). *Comp Biochem Physiol* 1:213–17.

Edwards, A.A., Mathura, C.B., Edwards, C.H. (1983). Effects of maternal geophagia on infant and juvenile rats. *J Nat Med Assoc* 75:895–902.

Edwards, C.H., Johnson, A.A., Knight, E.M., Oyemade, U.J., Cole, O.J., Westney, O.E., Jones, S., Laryea, H., Westney, L.S. (1994). Pica in an urban environment. *J Nutr* 124:954S–62S.

Eggenberger, M., Fluhmann, B., Muff, R., Lauber, M., Lichtensteeiger, W., Hunziker, W., Fischer, J.A., Born, W. (1996). Structure of a parathyroid hormone/parathyroid hormone-related peptide receptor of the human cerebellum and functional expression in human neuroblastoma SK-N-MC cells. *Mol Brain Res* 36:127–36.

Einarson, A., Koren, G., Bergman, U. (1998). Nausea and vomiting in pregnancy: a comparative European study. *Eur J Obstet Gynecol Reprod Biol* 76:1–3.

Ekman, P. (1982). *Emotion and the Human Face.* Cambridge University Press.

Elaroussi, M.A., DeLuca, H.F. (1994). Calcium uptake by chorioallantoic membrane: effects of vitamins D and K. *Am J Physiol* 267:E837–41.

Elaroussi, M.A., Forte, L.R., Eber, S.L., Biellier, H.V. (1994). Calcium homeostasis in the laying hen. 1. Age and dietary calcium effects. *Poult Sci* 873:1581–9.

Emmers, R., Nocenti, M.R. (1967). Role of thalamic gustatory nucleus in diet selection by normal and parathyroidectomized rats. *Proc Soc Exp Biol Med* 125:1264–70.

Epstein, A.N. (1982). Mineralocorticoids and cerebral angiotensin may act to produce sodium appetite. *Peptides* 3:493–4.

Epstein, A.N., Stellar, E. (1955). The control of salt preference in the adrenalectomized rat. *J Comp Physiol Psychol* 48:167–72.

Erskine, M.S. (1995). Prolactin release following mating and genitosensory stimulation in females. *Endocr Rev* 16:508–28.

Esque, T.C., Peters, E.L. (1992). Ingestion of bones, stones, and soil by desert tortoises. *Fish Wildlife Res* 13:105–11.

Evans, R.M. (1988). The steroid and thyroid hormone receptor superfamily. *Science* 240:889–95.

Evans, R.M., Arriza, J.L. (1989). A molecular framework for the actions of glucocorticoid hormones in the nervous system. *Neuron* 2:1105–12.

Evvard, J.M. (1915). Is the appetite of swine a reliable indication of physiological needs? *Proc Iowa Acad Sci* 22:375–403.

Fairweather-Tait, S.J., Johnson, A., Wright, A.J.A. (1993). The effect of dietary calcium intake in weanling rats on the efficiency of calcium absorption. *Br J Nutr* 69:527–32.

Fairweather-Tait, S., Prentice, A., Heumann, K.G., Jarjou, L.M., Stirling, L.M., Wharf, S.G., Turland, J.R. (1995). Effect of calcium supplements and stage of lactation on the calcium absorption efficiency of lactating women accustomed to low calcium intakes. *Am J Clin Nutr* 62:1188–92.

Fardellone, P., Sebert, J.L., Garabedian, M., Bellony, R., Maamer, M., Agbomson, F., Brazier, M. (1995). Prevalence and biological consequences of vitamin D deficiency in elderly institutionalized subjects. *Rev Rhum Engl Ed* 62:576–81.

Feinblatt, J.D., Tai, L.R., Kenny, A.D. (1975). Avian parathyroid glands in organ culture secretion of parathyroid hormone and calcitonin. *Endocrinology* 96:282–8.

Feldman, S.C., Christakos, S. (1983). Vitamin D-dependent calcium-binding protein in rat brain: biochemical and immunocytochemical characterization. *Endocrinology* 112:290–302.

Ferrante, B., Isa, L., Uderzo, A. (1993). Osservazioni sul metabolismo minerale osseo nella menopausa naturale e chirurgica. Ruolo della calcitonina sintetica di salmone e del calcio sul turnover osseo. (Observations on bone mineral metabolism in natural and surgical menopause. Role of synthetic salmon calcitonin and calcium on bone turnover). *Minerva Ginecol* 45:87–93.

Ferrell, F., Dreith, A.Z. (1986). Calcium appetite, blood pressure and electrolytes in spontaneously hypertensive rats. *Physiol Behav* 37:337–43.

Feskanich, D., Hunter, D.J., Willett, W.C., Hankinson, S.E., Hollis, B.W., Hough, H.L., Kelsey, K.T., Colditz, G.A. (1998). Vitamin D receptor genotype and the risk of bone fractures in women. *Epidemiology* 9:535–9.

Feskanich, D., Willet, W.C., Stampfer, M.J., Colditz, G.A. (1997). Milk, dietary calcium, and bone fractures in women: a 12-year prospective study. *Am J Public Health* 87:992–7.

Findlay, A.L., Fitzsimons, J.T., Kucharczyk, J. (1979). Dependence of spontaneous and angiotensin-induced drinking in the rat upon the oestrous cycle and ovarian hormones. *J Endocrinol* 82:215–25.

Fitzsimons, J.T. (1979). *The Physiology of Thirst and Sodium Appetite*. Cambridge University Press.

Fitzsimons, J.T. (1998). Angiotensin, thirst, and sodium appetite. *Physiol Rev* 78:583–686.

Fleming, A.S., Suh, E.J., Korsmit, M., Rusak, B. (1994). Activation of Fos-like immunoreactivity in the medial preoptic area and limbic structure by maternal and social interactions in rats. *Behav Neurosci* 108:724–34.

Flik, G., Fenwick, J.C., Kolar, Z., Mayer-Gostan, N., Wendelaar Bonga, S.E. (1986). Effects of ovine prolactin on calcium uptake and distribution in *Oreochromis mossambicus*. *Am J Physiol* 250:R161–6.

Flik, G., Fenwick, J.C., Wendelaar Bonga, S.E. (1989). Calcitropic actions of prolactin in freshwater North American eel (*Anguilla rostrata* LeSueur). *Am J Physiol* 257: R74–9.

Flik, G., Klaren, P.H.M., Schoenmakers, T.J.M., Bijvelds, M.J.C., Verbost, P.M., Bonga, S.E.W. (1995). Cellular calcium transport in fish: unique and universal mechanisms. *Physiol Zool* 69:403–17.

Fluharty, S.J., Epstein, A.N. (1983). Sodium appetite elicited by intracerebroventricular infusion of angiotensin II in the rat: II. Synergistic interaction with the systemic mineralocorticoids. *Behav Neurosci* 97:746–58.

Fluharty, S.J., Sakai, R.R. (1995). Behavioral and cellular studies of corticosterone and angiotensin interaction in brain. In: *Progress in Psychobiology and Physiological Psychology*, S.J. Fluharty, A.R. Morrison (eds.). San Diego: Academic Press.

Flynn, F.W., Gril, H.J., Schulkin, J., Norgren, R. (1992). Central gustatory lesions. II. Effects on sodium appetite, taste aversion learning, and feeding behaviors. *Behav Neurosci* 105:944–54.

Flynn, F.W., Schulkin, J., Havens, M. (1993). Sex differences in salt preference and taste reactivity in rats. *Brain Res Bull* 32:91–5.

Fox, N.A. (1985). Sweet/sour–interest/disgust: the role of approach–withdrawal in the development of emotions. In: *Infant Social Perception*, T. Field, N.A. Fox (eds.). Norwood, NJ: Ablex Press.

Frank, M.E. (1985). On the neural code for sweet and salty tastes. In: *Taste, Olfaction, and the CNS*, D.W. Pfaff (ed.). New York: Rockefeller University.

Frank, M.E. (1991). Taste-responsive neurons of the glossopharyngeal nerve of the rat. *J Neurophysiol* 65:1452–63.

Frank, M.E., Contreras, R.J., Hettinger, T.P. (1983). Nerve fibers sensitive to ionic taste stimuli in chorda tympani of the rat. *J Neurophysiol* 50:941–60.

Frankmann, S.P., Sakai, R.R., Simpson, J.B. (1987). Sodium appetite and cerebrospinal fluid sodium concentration during hypovolemia. *Appetite* 9:57–64.

Frankmann, S.P., Ulrich, P., Epstein, A.N. (1991). Transient and lasting effects of reproductive episodes on NaCl intake of the female rat. *Appetite* 16:193–204.

Fraser, D.R. (1980). Regulation of the metabolism of vitamin D. *Physiol Rev* 60:551–613.

Fraser, D.R. (1983). The physiological economy of vitamin D. *Lancet* 1:969–72.

Fraser, R.A., Zajac, J.D., Harvey, S. (1993). Expression of parathyroid hormone-related peptide gene in the rat hypothalamus. *Comp Biochem Physiol [B]* 106:647–50.

Fregly, M.J. (1974). Effect of an oral contraceptive on spontaneous activity of female rats. In: *Oral Contraceptives and High Blood Pressure*, M.J. Fregly, M.S. Fregly (eds.). Orlando: Dolphin Press.

Fregly, M.J. (1980). Effect of chronic treatment with estrogen on the dipsogenic response of rats to angiotensin. *Pharmacol Biochem Behav* 12:131–6.

Fregly, M.J., Thrasher, T.N. (1978). Attenuation of angiotensin-induced water intake in estrogen-treated rats. *Pharmacol Biochem Behav* 9:509–14.

Friedlander, G., Amiel, C. (1994). Cellular mode of action of parathyroid hormone. *Adv Nephrol Necker Hosp* 23:265–79.

Friedman, M.I., Bruno, J.P., Alberts, J.R. (1981). Physiological and behavioral consequences in rats of water recycling during lactation. *J Comp Physiol Psychol* 95:26–35.

Friedman, P.A., Gesek, F.A. (1993). Calcium transport in renal epithelial cells. *Am J Physiol* 264:F181–98.

Frye, C.A., DeMolar, G.L. (1994). Menstrual cycle and sex differences influence salt preference. *Physiol Behav* 55:193–7.

Fudim, O.K. (1978). Sensory preconditioning of flavors with a formalin-produced sodium need. *J Exp Psychol Anim Behav Proc* 4:276–85.

Fujita, T. (1996). Vitamin D in the treatment of osteoporosis revisited. *Proc Soc Exp Biol Med* 212:110–15.

Fukayama, S., Tashjian, A.H., Jr., Davis, J.N., Chisholm, J.C. (1995). Signaling by N- and C-terminal sequences of parathyroid hormone-related protein in hippocampal neurons. *Proc Natl Acad Sci USA* 92:10182–6.

Fuleihan, G.E.H., Brown, E.M., Gleason, R., Scott, J., Adler, G.K. (1996). Calcium modulation of adrenocorticotropin levels in women – a clinical research center study. *J Clin Endocrinol* 81:932–6.

Fuleihan, G.E.H., Klerman, E.B., Brown, E.N., Choe, Y., Brown, E.M., Czeisler, C.A. (1997). The parathyroid hormone circadian rhythm is truly endogenous – a general clinical research center study. *J Clin Endocrinol Metab* 82:281–6.

Fullmer, C.S. (1995). Dietary calcium levels and treatment interval determine the effects of lead ingestion on plasma 1,25-dihydroxyvitamin D concentration in chicks. *J Nutr* 125:1328–33.

Gaggi, R., DeIasio, R., Gianni, A.M. (1989). Relationships between the effects of peripherally administered salmon calcitonin on calcemia and brain biogenic amines. *Jpn J Pharmacol* 51:309–20.

Galaverna, O., DeLuca, L.A., Schulkin, J., Yao, S.Z., Epstein, A.N. (1992). Deficits in

NaCl ingestion after damage to the central nucleus of the amygdala in the rat. *Brain Res Bull* 28:89–98.

Galef, B.G., Jr. (1986). Social interaction modifies learned aversions, sodium appetite, and both palatability and handling time induced dietary preference in the rat. *J Comp Psychol* 100:432–9.

Galef, B.G., Jr. (1989). Enduring social enhancement of rats' preferences for the palatable and the piquant. *Appetite* 13:81–92.

Galef, B.G., Jr. (1991). A contrarian view of the wisdom of the body as it relates to dietary self-selection. *Psychol Rev* 98:218–23.

Galef, B.G., Jr. (1996). Food selection: problems in understanding how we choose foods to eat. *Neurosci Biobehav Rev* 20:67–73.

Galef, B.G., Jr. (1999). Is there a specific appetite for protein? In: *Neural and Metabolic Control of Macronutrient Intake*, H.R. Berthoud, R.J. Seeley (eds.). Boca Raton: CRC Press.

Gallagher, J.C., Goldgar, D., Moy, A. (1987). Total bone calcium in normal women: effect of age and menopause status. *J Bone Miner Res* 2:491–6.

Gallagher, J.C., Riggs, B.L., DeLuca, H.F. (1980). Effect of estrogen on calcium absorption and serum vitamin D metabolites in postmenopausal osteoporosis. *J Clin Endocrinol Metab* 51:1359–64.

Gallagher, S.J., Fraser, W.D., Owens, O.J., Dryburgh, F.J., Logue, F.C., Jenkins, A., Kennedy, J., Boyle, I.T. (1994). Changes in calciotrophic hormones and biochemical markers of bone turnover in normal human pregnancy. *Eur J Endocrinol* 131:369–74.

Gallin, W.J., Greenberg, M.E. (1995). Calcium regulation of gene expression in neurons: the mode of entry matters. *Curr Opin Neurobiol* 5:367–74.

Garcia, J., Ervin, F.R., Koelling, R.A. (1966). Learning with prolonged delay of reinforcement. *Psychonom Sci* 5:121–2.

Garcia, J., Hankins, W.G., Rusiniak, K.W. (1974). Behavioral regulation of the milieu interne in man and rat. *Science* 185:824–31.

Garcia, J., Kimmeldorf, D.J., Koelling, R.A. (1955). Conditioned aversion to saccharine resulting from exposure to gamma radiation. *Science* 122:157–8.

Garcia, J., Koelling, R.A. (1966). The relation of cue to consequence in avoidance learning. *Psychonom Sci* 5:123–4.

Garg, H.K., Singal, K.C., Arshad, Z. (1993). Zinc taste test in pregnant women and its correlation with serum zinc level. *Indian J Physiol Pharmacol* 37:318–22.

Garland, C.F., Garland, F.C., Gorham, E.D. (1991). Can colon cancer incidence and death rates be reduced with calcium and vitamin D? *Am J Clin Nutr* 54:193S–201S.

Gascon-Barre, M., Huet, P.M. (1983). Apparent [$^3$H]1,25-dihydroxyvitamin D uptake by canine and rodent brain. *Am J Physiol* 244:E266–71.

Gerardo-Gettens, T., Moore, B.J., Stern, J.S., Horwitz, B.A. (1989). Prolactin stimulates food intake in the absence of ovarian progesterone. *Am J Physiol* 256:R701–6.

Ghosh, A., Ginty, D.D., Bading, H., Greenberg, M.E. (1994). Calcium regulation of gene expression in neuronal cells. *J Neurobiol* 25:294–303.

Gibson, E.L., Booth, D.A. (1986). Acquired protein appetite in rats: dependence on a protein-specific need state. *Experientia* 42:1003–4.

Gibson, J.J. (1966). *The Senses Considered as Perceptual Systems*. Boston: Houghton Mifflin.

Glorieux, F.H., St. Arnaud, R. (1998). Molecular cloning of (25-OH D)-1-alpha-hydroxylase: an approach to the understanding of vitamin D pseudo-deficiency. *Recent Prog Horm Res* 53:341–9.

Gloth, F.M., III, Gundberg, C.M., Hollis, B.W., Haddad, J.G., Jr., Tobin, J.D. (1995). Vitamin D deficiency in homebound elderly persons. *JAMA* 274:1683–6.

Goldberg, G.R., Prentice, A.M. (1994). Maternal and fetal determinants of adult disease. *Nutr Rev* 52:191–200.

Goldberg, G.R., Prentice, A.M., Coward, W.A., Davies, H.L., Murgatroyd, P.R., Wensing, C., Black, A.E., Harding, M., Sawyer, M. (1993). Longitudinal assessment of energy expenditure in pregnancy by the doubly labeled water method. *Am J Clin Nutr* 57:494–505.

Goldsmith, N.J., Johnson, J.O. (1975). Bone mineral: effects of oral contraceptives, pregnancy, and lactation. *J Bone Joint Surg* 57A:657 (abstract).

Gompel, A., Truc, J.B., Decroix, Y., Poitout, P. (1993). *Place du calcium aux différents ages de la vie des femmes.* Paris: La Presse Médicale Masson.

Gona, A.G., Pendurthi, T.K., Al-Rabiai, S., Gona, O., Christakas, S. (1986). Immunocytochemical localization and immunological characterization of vitamin D-dependent calcium-binding protein in the bullfrog cerebellum. *Brain Behav Evol* 29:176–83.

Goodall, E.M., Whittle, M., Cookson, J., Cowen, P.J., Silverstone, T. (1995). Menstrual cycle effects on the action of buspirone on food intake in healthy female volunteers. *J Psychopharmacol* 9:307–12.

Goodall, J. (1986). *The Chimpanzee of the Gombe.* Cambridge, MA: Harvard University Press.

Gorski, R.A., Gordon, J.H., Shryne, J.E., Southam, A.M. (1978). Evidence for a morphological sex difference within the medial preoptic area of the rat brain. *Brain Res* 148:333–46.

Gould, S.J. (1977). *Ontogeny and Phylogeny.* Cambridge, MA: Harvard University Press.

Gould, S.J., Eldridge, N. (1977). Punctuated equilibria: the tempo and mode of evolution considered. *Paleobiology* 3:115–51.

Goy, R.W., McEwen, B.S. (1977). *Sexual Differentiation of the Brain.* Cambridge, MA: MIT Press.

Green, H.G. (1925). Perverted appetites. *Physiol Rev* 5:336–45.

Green, J., McIntosh, M., Wilson, A. (1991). Changes in nutrition knowledge scores and calcium intake in female adolescents. *Home Econ Res J* 19:207–14.

Greer, F.R. (1989). Calcium, phosphorus, and magnesium: how much is too much for infant formulas? *J Nutr* 119:1846–51.

Greer, F.R., Lane, J., Ho, M. (1984). Elevated serum parathyroid hormone, calcitonin, and 1,25-dihydroxyvitamin D in lactating women nursing twins. *Am J Clin Nutr* 40:562–8.

Greer, F.R., Searcy, J.E., Levin, R.S., Steichen, J.J., Steichen-Asche, P.S., Tsang, R.C. (1982a). Bone mineral content and serum 25-hydroxyvitamin D concentrations in breast-fed infants with and without supplemental vitamin D: one-year follow-up. *J Pediatr* 100:919–22.

Greer, F.R., Steichen, J.J., Tsang, R.C. (1982b). Effects of increased calcium, phosphorus, and vitamin D intake on bone mineralization in very low-birth-weight infants fed formulas with polycose and medium-chain triglycerides. *J Pediatr* 100:951–5.

Greer, F.R., Tsang, R.C., Searcy, J.E., Levin, R.S., Steichen, J.J. (1982c). Mineral homeostasis during lactation – relationship to serum 1,25-dihydroxyvitamin D, 25-hydroxyvitamin D, parathyroid hormone, and calcitonin. *Am J Clin Nutr* 36:431–7.

Greger, J.L., Baliger, P., Abernathy, R.P., Bennett, O.A., Peterson, T. (1978). Calcium, magnesium, phosphorus, copper, and manganese balance in adolescent females. *Am J Clin Nutr* 31:117–21.

Grigsby, R.K., Thyer, B.A., Waller, R.J., Johnston, G.A., Jr. (1999). Chalk eating in middle Georgia: a culture-bound syndrome of pica? *South Med J* 92:190–2.

Grigson, P.S., Kaplan, J.M., Roitman, M.F., Norgren, R., Grill, H.J. (1997). Reward comparison in chronic decerebrate rats. *Am J Physiol* 273:R479–86.

Grill, H.J. (1980). Production and regulation of consummatory behavior in the chronic decerebrate rat. *Brain Res Bull* 5:79–87.

Grill, H.J., Berridge, K.C. (1985). Taste reactivity as a measure of the neural control of palatability. In: *Progress in Psychobiology and Physiological Psychology*, A.N. Epstein, S.J. Sprague (eds.). San Diego: Academic Press.

Grill, H.J., Norgren, R. (1978). The taste reactivity test. II. Mimetic responses to gustatory stimuli in chronic thalamic and chronic decerebrate rats. *Brain Res* 143:281–97.

Grill, H.J., Schulkin, J., Flynn, F.W. (1986). Sodium homeostasis in chronic decerebrate rats. *Behav Neurosci* 100:536–43.

Grillo, C.A., Coirini, H., McEwen, B.S., DeNicola, A.F. (1989). Changes of salt intake and of (Na + K)-ATPase activity in brain after high dose treatment with deoxycorticosterone. *Brain Res* 499:225–33.

Grillo, C.A., Saravia, F., Ferrini, M., Piroli, G., Roig, P., Garcia, S.I., de Kloet, E.R., DeNicola, A.F. (1998). Increased expression of magnocellular vasopressin mRNA in rats with deoxycorticosterone-acetate induced salt appetite. *Neuroendocrinology* 68:105–15.

Grodstein, F., Stampfer, M.J., Colditz, G.A., Willett, W.C., Manson, J.E., Joffe, M., Rosner, B., Fuchs, C., Hankinson, S.E., Hunter, D.J., Hennekens, C.H., Speizer, F.E. (1997). Postmenopausal hormone therapy and mortality. *N Eng J Med* 336:1769–75.

Gubernick, D.J., Nelson, R.J. (1989). Prolactin and paternal behavior in the biparental California mouse, *Peromyscus californicus*. *Horm Behav* 23:203–10.

Guidobono, F., Netti, C., Pecili, A., Gritti, I., Mancia, M. (1987). Calcitonin binding site distribution in the cat central nervous system: a wider insight of the peptide involvement in brain functions. *Neuropeptides* 10:265–73.

Guisado, R., Arieff, A.I., Massry, S.G., Lazarowitz, V., Kerian, A. (1975). Changes in the electroencephalogram in acute uremia. Effects of parathyroid hormone and brain electrolytes. *J Clin Invest* 55:738–45.

Gustafson, E.L., Greengard, P. (1990). Localization of DARPP-32 immunoreactive neurons in the bed nucleus of the stria terminalis and central nucleus of the stria terminalis and central nucleus of the amygdala: co-distribution with axons containing tyrosine hydroxylase, vasoactive intestinal polypeptide and calcitonin gene-related peptide. *Exp Brain Res* 79:447–58.

Hacking, I. (1975). *The Emergence of Probability*. Cambridge University Press.

Hainsworth, F.R., Wolf, L.L. (1990). Comparative studies of feeding. Chapter 11 in *Handbook of Behavioral Neurobiology. Vol. 10: Neurobiology of Food and Fluid Intake*, E.M. Stricker (ed.). New York: Plenum Press.

Hall, A.K., Norman, A.W. (1991). Vitamin D-independent expression of chick brain calbindin-D28K. *Brain Res Mol Brain Res* 9:9–14.

Hall, W.G. (1979). Feeding and behavioral activation in infant rats. *Science* 205:206–9.

Halloran, B.P., DeLuca, H.F. (1980). Calcium transport in small intestine during early development: role of vitamin D. *Am J Physiol* 239:G473–9.

Halperin, M.L., Kamel, K.S. (1998). Potassium. *Lancet* 352:135–40.

Haramati, A., Haas, J.A., Knox, F.G. (1983). Adaptation of deep and superficial nephons to changes in dietary phosphate intake. *Am J Physiol* 244:F265–9.

Haramati, A., Mulroney, S.E., Webster, S.K. (1988). Developmental changes in the tubular capacity for phosphate reabsorption in the rat. *Am J Physiol* 255:F287–91.

Harriman, A. (1955). Provitamin A selection by vitamin A depleted rats. *J Physiol* 86:45–50.

Harris, G.W. (1964). Sex hormones, brain development and brain function. *Endocrinology* 75:627–48.

Harris, L.J., Clay, J., Harvreves, F., Ward, A. (1933). Appetite and choice of diet. The ability of the vitamin B deficient rat to discriminate between diets containing and lacking the vitamin. *Proc R Soc Lond B* 113:161–90.

Harris, R.A., Carnes, D.L., Forte, L.R. (1981). Reduction of brain calcium after consumption of diets deficient in calcium or vitamin D. *J Neurochem* 36:460–6.

Harris, S.S., Dawson-Hughes, B. (1994). Caffeine and bone loss in healthy postmenopausal women. *Am J Clin Nutr* 60:573–8.

Hartard, M., Bottermann, P., Bartenstein, P., Jeschke, D., Schwaiger, M. (1997). Effects on bone mineral density of low-dosed oral contraceptives compared to and combined with physical activity. *Contraception* 55:87–90.

Harvey, S., Hayer, S. (1993). Parathyroid hormone binding sites in the brain. *Peptides* 14:1187–91.

Hastings, J.J., Christian, D.P., Manning, T.E., Harth, C.C. (1991). Sodium and potassium effects on adrenal-gland indices of mineral balance in meadow voles. *J Mammol* 72:641–51.

Hatton, D.C., Yue, Q., McCarron, D.A. (1995). Mechanisms of calcium' s effects on blood pressure. *Semin Nephrol* 15:593–7.

Hayslip, C.C., Klein, T.A., Wray, H.L., Duncan, W.E. (1989). The effects of lactation on bone mineral content in healthy postpartum women. *Obstet Gynecol* 73:588–92.

He, T.-C., Sparks, A.B., Rago, C., Hermeking, H., Zawel, L., da Costa, L.T., Morin, P.J., Vogelstein, B., Kinzler, K.W. (1998). Identification of c-MYC as a target of the APC pathway. *Science* 281:1509–12.

Heaney, R.P. (1991). Calcium intake in the osteoporotic fracture context: introduction. *Am J Clin Nutr* 54:242S–4S.

Heaney, R.P. (1992). Calcium in the prevention and treatment of osteoporosis. *J Intern Med* 231:169–80.

Heaney, R.P. (1993). Nutritional factors in osteoporosis. *Annu Rev Nutr* 13:187–216.

Heaney, R.P. (1996). Pathophysiology of osteoporosis. *Am J Med Sci* 312:251–6.

Heaney, R.P., Barger-Lux, M.J. (1994). ADSA Foundation lecture. Low calcium intake: the culprit in many chronic diseases. *J Dairy Sci* 77:1155–60.

Heaney, R.P., Recker, R.R. (1982). Effects of nitrogen, phosphorus, and caffeine on calcium balance in women. *J Lab Clin Med* 99:46–55.

Heaney, R.P., Recker, R.R. (1994). Determinants of endogenous fecal calcium in healthy women. *J Bone Miner Res* 9:1621–7.

Heaney, R.P., Recker, R.R., Hinders, S.M. (1988). Variability of calcium absorption. *Am J Clin Nutr* 47:262–4.

Heaney, R.P., Reker, R.R., Saville, P.D. (1977). Calcium balance and calcium requirements in middle-aged women. *Am J Clin Nutr* 30:1603–11.

Heaney, R.P., Reker, R.R., Saville, P.D. (1978). Menopausal changes in calcium balance performance. *J Lab Clin Med* 92:953–63.

Heaney, R.P., Recker, R.R., Weaver, C.M. (1990). Absorbability of calcium sources: the limited role of solubility. *Calcif Tissue Int* 46:300–4.

Heaney, R.P., Saville, P.D., Recker, R.R. (1975). Calcium absorption as a function of calcium intake. *J Lab Clin Med* 85:881–90.

Heaney, R.P., Skillman, T.G. (1971). Calcium metabolism in normal human pregnancy. *J Clin Endocrinol Metab* 33:661–70.

Heaney, R.P., Weaver, C.M. (1990). Calcium absorbtion from kale. *Am J Clin Nutr* 51:656–7.

Hearn, J.P. (1983). The common marmoset. In: *Reproduction in New World Primates*, J.P. Hearn (ed.). Lancaster, PA: MTP Press.

Hebert, D., Cowan, I.M. (1971). Natural salt licks as a part of the ecology of the mountain goat. *Can J Zool* 49:605–10.

Hebert, S.C. Brown, E.M. (1995). The extracellular calcium receptor. *Curr Opin Cell Biol* 7:484–92.

Hebert, S.C., Brown, E.M., Harris, H.W. (1997). Role of the Ca(2+)-sensing receptor in divalent mineral ion homeostasis. *J Exp Biol* 200:295–302.

Heini, A., Shutz, Y., Diaz, E., Prentice, A.M., Whitehead, R.G., Jequier, E. (1991). Free-living energy expenditure measured by two independent techniques in pregnant and nonpregnant Gambian women. *Am J Physiol* 261:E9–17.

Hellwald, H. (1931). Untersuchungen über Tribstarken bei Tieren. *Z Psychol* 123:91–141.

Herbert, J. (1993). Peptides in the limbic system: neurochemical codes or co-ordinated adaptive responses to behavioural and physiological demand. *Neurobiology* 41:723–91.

Herrera, J.A., Arevalo-Herrera, M., Herrera, S. (1998). Prevention of preeclampsia by linoleic acid and calcium supplementation: a randomized controlled trial. *Obstet Gynecol* 91:585–90.

Herrick, C.J. (1905). The central gustatory paths in the brains of bony fishes. *J Comp Neurol Psychol* 15:375–456.

Herrick, C.J. (1918). *An Introduction to Neurology*. Philadelphia: Saunders.

Herrick, C.J. (1948). *The Brain of the Tiger Salamander*. University of Chicago Press.

Hilakivi-Clarke, L. (1994). Overexpression of transforming growth factor alpha in transgenic mice alters nonreproductive, sex-related behavioral differences; interaction with gonadal hormones. *Behav Neurosci* 108:410–17.

Hill, D.L., Formaker, B.K., White, K.S. (1990). Perceptual characteristics of the amiloride-suppressed sodium chloride taste response in the rat. *Behav Neurosci* 104:734–41.

Hill, D.L., Mistretta, C.M. (1990). Developmental neurobiology of salt taste sensation. *Trends Neurosci* 13:188–95.

Hilton, J.M., Chai, S.Y., Sexton, P.M. (1995). In vitro autoradiographic localization of the calcitonin receptor isoforms, Cla and Clb, in rat brain. *Neuroscience* 69:1223–37.

Hoare, S., Poppitt, S.D., Prentice, A.M., Weaver, L.T. (1996). Dietary supplementation and rapid catch-up growth after acute diarrhoea in childhood. *Br J Nutr* 76:479–90.

Hoebel, B.G. (1997). Neuroscience and appetitive behavior research: 25 years. *Appetite* 29:119–33.

Hoener, R.D., Lackey, C.J., Kolasa, K., Warren, K. (1991). Pica practices of pregnant women. *J Am Diet Assoc* 91:34–8.

Hoffman, R.A., Robinson, P.F. (1966). Changes in some endocrine glands of white-tailed deer as affected by season, sex, and age. *J Mammol* 47:266–80.

Hoffman, S., Grisso, J.A., Kelsey, J.L., Gammon, M.D., O'Brien, L.A. (1993). Parity, lactation and hip fracture. *Osteoporos Int* 3:171–6.

Hogan-Warburg, A.J., Hogan, J.A. (1981). Feeding strategies in the development of food recognition in young chicks. *Anim Behav* 29:143–54.

Holick, M.F. (1986). Vitamin D requirements for the elderly. *Clin Nutr* 5:121–9.

Holick, M.F. (1994). Vitamin D – new horizons for the 21st century. *Am J Clin Nutr* 60:619–30.

Holick, M.F., Matsuoka, L.Y., Wortsman, J. (1989). Age, vitamin D, and solar radiation. *Lancet* 2:1104–5.

Hollis, B.W. (1996). Assessment of vitamin D nutritional and hormonal status: what to measure and how to do it. *Calcif Tissue Int* 58:4–5.

Holman, E. (1975). Immediate and delayed reinforcers for flavor preferences in rats. *Learn Motiv* 6:91–100.

Holmes, R.P., Kummeroa, F.A. (1983). The necessity and safety of calcium and vitamin D in the elderly. *J Am Coll Nutr* 2:173–99.

Holt, E.H., Broadus, A.E., Brinex, M.L. (1996). Parathyroid hormone-related pepide is produced by cultured cerebellar granule cells in response to L-type voltage-sensitive $Ca^{2+}$channel flux via a $Ca^{2+}$/calmodulin-dependent kinase pathway. *J Biol Chem* 271:28105–11.

Holt, P.R., Atillasoy, E.O., Gilman, J., Guss, J., Moss, S.F., Newmark, H., Fan, K., Yang, K., Lipkin, M. (1998). Modulation of abnormal colonic epithelial cell proliferation and differentiation by low-fat dairy foods: a randomized controlled trial. *JAMA* 280:1074–9.

Hood, W.R. (1998). Meeting the nutritional demands of lactation: calcium constraints in an insectivorous bat. *Proc Comp Nutr Soc* 2:92–6.

Hook, E.B. (1978). Dietary cravings and aversions during pregnancy. *J Am Geriatr Soc* 38:862–6.

Hosking, D.J. (1996). Calcium homeostasis in pregnancy. *Clin Endocrinol* 45:1–6.

Hotchkiss, C.E., Jerome, C.P. (1998). Evaluation of a nonhuman primate model to study circadian rhythms of calcium metabolism. *Am J Physiol* 275:R494–501.

Hughes, B.O. (1972). A circadian rhythm of calcium intake in the domestic fowl. *Br Poultry Science* 13:485–93.

Hughes, B.O., Wood-Gush, D.G. (1971). A specific appetite for calcium in domestic chickens. *Anim Behav* 19:490–9.

Hughes, B.O., Wood-Gush, D.G.M. (1972). Hypothetical mechanisms underlying calcium appetite in fowls. *Rev Comp Animal* 14:95–106.

Hunscher, H.A. (1930). Metabolism of women during the reproductive cycle. II. Calcium and phosphorus utilization in two successive lactation periods. *J Biol Chem* 86:37–57.

Hutton, L.A., Gu, G., Simerly, R.B. (1998). Development of a sexually dimorphic projection from the bed nuclei of the stria terminalis to the anteroventral periventricular nucleus in the rat. *J Neurosci* 18:3003–13.

Iacopino, A.M., Christakos, S. (1990). Corticosterone regulates calbindin-D mRNA and protein levels in rat hippocampus. *J Biol Chem* 265:10177–80.

Iida, K., Shinki, T., Yamaguchi, A., DeLuca, H.F., Kurokawa, K., Suda, T. (1995). A possible role of vitamin D receptors in regulationg vitamin D activation in the kidney. *Proc Natl Acad Sci USA* 92:6112–16.

Imai, T. (1995). The effects of long-term intake of restricted calcium, vitamin D, and vitamin E and cadmium-added diets on the bone mass of mouse femoral bone: a microdensitometrical study (in Japanese). *Nippon Eiseigaku Zasshi* 50:754–62.

Ingram, M.C., Wallace, A.M., Collier, A., Fraser, R., Connell, J.M. (1996). Sodium status, corticosteroid metabolism and blood pressure in normal human subjects and in a patient with abnormal salt appetite. *Clin Exp Pharmacol Physiol* 23:375–8.

Inoue, M., Tordoff, M.G. (1998). Calcium deficiency alters chorda tympani nerve responses to oral calcium chloride. *Physiol Behav* 63:297–303.

Institute of Medicine. (1997). *Dietary Reference Intakes. Calcium, Phosphorus, Magnesium, Vitamin D, and Fluoride*. Washington, DC: National Academy Press.

Inzucchi, S.E., Robbins, R.J. (1994). Effects of growth hormone on human bone biology. *J Clin Endocrinol Metab* 7:691–4.

Irving, J.T. (1957). *Calcium Metabolism*. London: Methuen.

Jackson, J.H. (1958). The Croonian Lectures on evolution and dissolution of the nervous system. In: *Selected Writings of John Hughlings Jackson*, J. Taylor (ed.). London: Staples Press. (Reprinted from *Br Med J*, 1884, I, 591–3, 660–3, 703–7.)

Jacobowitz, D.M., Winsky, L. (1991). Immunocytochemical localization of calretinin in the forebrain of the rat. *J Comp Neurol* 304:198–218.

Jacobson, J.L., Snowdon, C.T. (1976). Increased lead ingestion in calcium-deficient monkeys. *Nature* 262:51–2.

Jacoby, I., Meyer, G.S., Haffner, W., Cheng, E.Y., Potter, A.L., Pearse, W.H. (1998). Modeling the future workforce of obstetrics and gynecology. *Obstet Gynecol* 92: 450–6.

Jakinovich, W., Jr., Osborn, D.W. (1981). Zinc nutrition and salt preference in rats. *Am J Physiol* 241:R233–9.

Jande, S.S., Maler, L., Lawson, D.E.M. (1981). *Immunohistochemical Mapping of Vitamin D-dependent Calcium-binding Protein in Brain*. London: Macmillan Journals Ltd.

Janzen, D.H. (1977). Why fruit rots, seeds mold and meat spoils. *Amer Nat* 111:691–713.

Jehan, F., DeLuca, H.F. (1997). Cloning and characterization of the mouse vitamin D receptor promoter. *Proc Natl Acad Sci USA* 94:10138–43.

Jehan, F., Ismail, R., Hanson, K., DeLuca, H.F. (1998). Cloning and expression of the chicken 25-hydroxyvitamin $D_3$ 24-hydroxylase cDNA. *Biochim Biophys Acta* 1395: 259–65.

Joborn, C., Hetta, J., Niklasson, F., Rastad, J., Wide, L., Agren, H., Akerstrom, G., Ljunghall, S. (1991). Cerebrospinal fluid calcium, parathyroid hormone, and monoamine and purine metabolites and the blood-brain barrier function in primary hyperparathyroidism. *Psychoneuroendocrinology* 16:311–22.

Johnson, A.K., Thunhorst, R.L. (1997). The neuroendocrinology of thirst and salt appetite: visceral sensory signals and mechanisms of central integration. *Front Neuroendocrinol* 18:292–353.

Johnson, C.M., Hill, C.S., Chawla, S., Treisman, R., Bading, H. (1997). Calcium controls gene expression via three distinct pathways that can function independently of the Ras/mitogen-activated protein kinases (ERKs) signaling cascade. *J Neurosci* 17:6189–202.

Johnson, J.A., Beckman, M.J., Pansini-Porta, A., Christakos, S., Bruns, M.E., Beitz, D.C., Horst, R.L., Reinhardt, T.A. (1995). Age and gender effects on 1,25-dihydroxyvitamin D–regulated gene expression. *Exp Gerontol* 30:631–43.

Johnson, K.R., Jobber, J., Stonawski, B.J. (1980). Prophylactic vitamin D in the elderly. *Age Ageing* 9:121–7.

Jones, G., Strugnell, S.A., DeLuca, H.F. (1998). Current understanding of the molecular actions of vitamin D. *Physiol Rev* 78:1193–231.

Jones, R.L., Hanson, H.C. (1985). *Mineral Licks: Geography and Biogeochemistry of North American Ungulates*. Ames: Iowa State University Press.

Jonklaas, J., Buggy, J. (1985a). Angiotensin–estrogen interaction in female brain reduces drinking and pressor responses. *Am J Physiol* 247:R167–72.

Jonklaas, J., Buggy, J. (1985b). Angiotensin–estrogen central interaction: localization and mechanism. *Brain Res* 326:239–49.

Joshua, I.G., Mueller, W.J. (1979). The development of a specific appetite for calcium in growing broiler chicks. *Br Poult Sci* 20:481–90.

Jovanovic-Peterson, L., Peterson, C.M. (1996). Vitamin and mineral deficiencies which may predispose to glucose intolerance of pregnancy. *J Am Coll Nutr* 15:14–20.

Kalkwarf, H.J., Specker, B.L. (1995). Bone mineral loss during lactation and recovery after weaning. *Obstet Gynecol* 86:26–32.

Kalkwarf, H.J., Specker, B.L., Bianchi, D.C., Ranz, J., Ho, M. (1997). The effect of calcium supplementation on bone density during lactation and after weaning. *N Engl J Med* 337:523–8.

Kalkwarf, H.J., Specker, B.L., Heubi, J.E., Vieira, N.E., Yergey, A.L. (1996). Intestinal calcium absorption of women during lactation and after weaning. *Am J Clin Nutr* 63:526–31.

Kalkwarf, H.J., Specker, B.L., Ho, M. (1999). Effects of calcium supplementation on calcium homeostasis and bone turnover in lactating women. *J Clin Endocrinol Metab* 84:464–70.

Kanarek, R.B., Ryu, M., Przypek, J. (1995). Preferences for foods with varying levels of salt and fat differ as a function of dietary restraint and exercise but not menstrual cycle. *Physiol Behav* 57:821–6.

Kaneko, T., Pang, P.K. (1987). Immunocytochemical detection of parathyroid hormone-like substance in the goldfish brain and pituitary gland. *Gen Comp Endocrinol* 68: 147–52.

Kanis, J.A. (1994). Calcium nutrition and its implications for osteoporosis. Part I. Children and healthy adults. *Eur J Clin Nutr* 48:757–67.

Kanis, J.A., Pitt, F.A. (1992). Epidemiology of osteoporosis. *Bone* 13:S7–15.

Karaplis, A.C., Luz, A., Glowacki, J., Bronson, R.T., Tybulewicz, V.L., Kronenberg, H.M., Mulligan, R.C. (1994). Lethal skeletal dysplasia from targeted disruption of the parathyroid hormone-related peptide gene. *Genes Dev* 8:277–89.

Karkoszka, H., Chudek, J., Strzelczyk, P., Wiecek, A., Schmidt-Gayk, H., Ritz, E., Kokot, F. (1998). Does the vitamin D receptor genotype predict bone mineral loss in haemodialysed patients? *Nephrol Dial Transplant* 13:2077–80.

Karmali, R., Schiffmann, S.N., Vanderwinden, J.M., Hendy, G.N., Nys-DeWolf, N., Corvilain, J., Bergmann, P., Vanderhaeghen, J.J. (1992). Expression of mRNA of parathyroid hormone-related peptide in fetal bones of the rat. *Cell Tissue Res* 270:597–600.

Karoui, A., Karoui, H. (1993). Pica in Tunisian children. Results of a survey performed in a polyclinic of the Tunisian social security national administration (in French). *Pédiatrie* 48:565–9.

Karpf, D.B., Shapiro, D.R., Seeman, E., Ensrud, K.E., Johnston, C.C., Adami, S., Harris, S.T., Santora, A.C., Hirsch, L.J., Oppenheimer, L., Thompson, D. (1997). Prevention of nonvertebral fractures by alendronate. A meta-analysis. Alendronate osteoporosis treatment study group. *JAMA* 277(14):1159–64.

Katz, D. (1937). *Animals and Men: Studies in Comparative Psychology.* London: Longman.

Katz, S.H., Foulks, E.F. (1970). Mineral metabolism and behavior: abnormalities of calcium homeostasis. *Am J Phys Anthropol* 32:299–304.

Kaufman, P. (1989). Cultural aspects of nutrition. *Top Clin Nutr* 4:1–6.

Kaufman, S. (1980). A comparison of the dipsogenic responses of male and female rats to a variety of stimuli. *Can J Physiol Pharmacol* 58:1180–3.

Kaufman, S. (1981). Control of fluid intake in pregnant and lactating rats. *J Physiol* 318: 9–16.

Kaufman, S., Mackay, B.J. (1983). Plasma prolactin levels and body fluid deficits in the rat: causal interactions and control of water intake. *J Physiol* 336:73–81.

Kaufman, S., Mackay, B.J., Scott, J.Z. (1981). Daily water and electrolyte balance in chronically hyperprolactinaemic rats. *J Physiol* 321:11–19.

Keeler, J.O., Studier, E.H. (1992). Nutrition in pregnant big brown bats (*Eptesicus fuscus*) feeding on June beetles. *J Mammol* 73:426–30.

Keith, D., Keith, L., Berger, G.S., Foot, J., Webster, A. (1975). Amylophagia during pregnancy: some maternal and perinatal correlations. *Mt Sinai J Med* 42:410–14.

Kelsay, J.L., Prather, E.S. (1983). Mineral balances of human subjects consuming spinach in a low-fiber diet and in a diet containing fruits and vegetables. *Am J Clin Nutr* 38:12–19.

Kent, G.N., Price, R.I., Gutteridge, D.H., Allen, J.R., Rosman, K.J., Smith, M., Bhagat, C.I., Wilson, S.G., Retallack, R.W. (1993). Effect of pregnancy and lactation on maternal bone mass and calcium metabolism. *Osteoporos Int [Suppl 1]* 3:44–7.

Kent, G.N., Price, R.I., Gutteridge, D.H., Rosman, K.J., Smith, M., Allen, J.R., Hickling, C.J., Blackeman, S.L. (1991). The efficiency of intestinal calcium absorption is increased in late pregnancy but not in established lactation. *Calcif Tissue Int* 48: 293–5.

Kent, G.N., Price, R.I., Gutteridge, D.H., Smith, M., Allen, J.R., Bhagat, C.I., Barnes, M.P., Hickling, C.J., Retallack, R.W., Wilson, S.G., et al. (1990). Human lactation: forearm trabecular bone loss, increased bone turnover, and renal conservation of calcium and inorganic phosphate with recovery of bone mass following weaning. *J Bone Miner Res* 5:361–9.

Kerstetter, J.E., O'Brien, K.O., Insogna, K.L. (1998). Dietary protein affects intestinal calcium absorption. *Am J Clin Nutr* 68:859–65.

Khudaverdyan, D.N., Ter-Markosian, A.S., Tadevosyan, Y.V. (1997). Initiation of the phosphoinositide cycle by parathyroid hormone in synaptosomes. *Biochemistry (Moscow)* 62:1109–12.

Kilby, M.D., Broughton, P.F., Symonds, E.M. (1993). Changes in platelet intracellular free calcium in normal pregnancy. *Br J Obstet Gynaecol* 100:375–9.

Kimmel-Jehan, C., Jehan, F., DeLuca, H.F. (1997). Salt concentration determines 1,25-dihydroxyvitamin $D_3$ dependency of vitamin D receptor–retinoid X receptor–vitamin D-responsive element complex formation. *Arch Biochem Biophys* 341:75–80.

Kimura, D. (1995). Estrogen replacement therapy may protect against intellectual decline in postmenopausal women. *Horm Behav* 29:312–21.

King, C.T., Hill, D.L. (1993). Neuroanatomical alterations in the rat nucleus of the solitary tract following early maternal NaCl repletion. *J Comp Neurol* 333:531–2.

Kinyamu, H.K., Gallagher, J.C., Knezetic, J.A., DeLuca, H.F., Prahl, J.M., Lanspa, S.J. (1997a). Effect of vitamin D receptor genotypes on calcium absorption, duodenal vitamin D receptor concentration, and serum 1,25-dihydroxyvitamin D levels in normal women. *Calcif Tissue Int* 60:491–5.

Kinyamu, H.K., Gallagher, J.C., Prahl, J.M., DeLuca, H.F., Petranick, K.M., Lanspa, S.J. (1997b). Association between intestinal vitamin D receptor, calcium absorption, and serum 1,25-dihydroxyvitamin D in normal young and elderly women. *J Bone Miner Res* 12:922–8.

Kirksey, A., Ernst, J.A., Roepke, J.L., Tsai, T.L. (1979). Influence of mineral intake and use of oral contraceptives before pregnancy on the mineral content of human colostrum and of more mature milk. *Am J Clin Nutr* 32:30–9.

Kisley, L.R., Sakai, R.R., Ma, L.Y., Fluharty, S.J. (1999). Ovarian steroid regulation of angiotensin II-induced water intake in rats. *Am J Physiol* 276:R90–6.

Klein, C.J., Moser-Veillon, P.B., Douglass, L.W., Ruben, K.A., Trocki, O. (1995). A longitudinal study of urinary calcium, magnesium, and zinc excretion in lactating and nonlactating postpartum women. *Am J Clin Nutr* 61:779–86.

Kloub, M.A., Heck, G.L., DeSimone, J.A. (1998). Self-inhibition in $Ca^{2+}$-evoked taste responses: a novel tool for functional dissection of salt taste transduction mechanisms. *J Neurophysiol* 79:911–21.

Knox, B., Kremeer, J., Pearce, J. (1995). A survey of dietary urges and consumption during pregnancy in Belfast working-class women. *Soc Sci Health* 1:125–44.

Kochersberger, G., Westlund, R., Lyles, K.W. (1991). The metabolic effects of calcium supplementation in the elderly. *J Am Geriatr Soc* 39:192–6.

Koetting, C.A., Wardlaw, G.M. (1988). Wrist, spine, and hip bone density in women with variable histories of lactation. *Am J Clin Nutr* 48:1479–81.

Kon, S.K. (1931).The self-selection of food constituents by the rat. *Biochem J* 25:473–81.

Korman, S.H. (1990). Pica as a presenting symptom in childhood celiac disease. *Am J Clin Nutr* 51:139–41.

Kovacs, C.S., Chik, C.L. (1995). Hyperprolactinemia caused by lactation and pituitary adenomas is associated with altered serum calcium, phosphate, parathyroid hormone (PTH), and PTH-related peptide levels. *J Clin Endocrinol Metab* 80:3036–42.

Kovacs, C.S., Kronenberg, H.M. (1997). Maternal-fetal calcium and bone metabolism during pregnancy, puerperium, and lactation. *Endocr Rev* 18:832–72.

Krall, E.A., Sahyoun, N., Tannenbaum, S., Dallal, G.E., Dawson-Hughes, B. (1989). Effect of vitamin D intake on seasonal variations in parathyroid hormone secretion in postmenopausal women. *N Engl J Med* 321:1777–83.

Krebs, J.R., Davies, N.B. (eds.) (1984). *Behavioral Ecology*, 2nd ed. Oxford: Blackwell.

Krebs, N.F., Reidinger, C.J., Robertson, A.D., Brenner, M. (1997). Bone mineral density changes during lactation: maternal, dietary, and biochemical correlates. *Am J Clin Nutr* 65:1738–46.

Krecek, J. (1978). Effect of ovariectomy of females and oestrogen administration to males during the neonatal critical period on salt intake in adulthood in rats. *Physiol Bohemoslov* 27:1–5.

Krecek, J., Novakova, V., Panek, M.M., Salatova, J., Sterc, J., Zicha, J. (1972a). Critical periods of development and the regulation of salt intake. *Neuroontogenti*, pp. 249–58.

Krecek, J., Novakova, V., Stibral, K. (1972b). Sex differences in the taste preference for a salt solution in the rat. *Physiol Behav* 8:183–8.

Krieckhaus, E.E. (1970). "Innate recognition" aids rats in sodium regulation. *J Comp Physiol Psychol* 75:117–22.

Krieckhaus, E.E., Wolf, G. (1968). Acquisition of sodium by rats: interaction of innate mechanisms and learning. *J Comp Physiol Psychol* 65:197–201.

Krolner, B. (1983). Seasonal variation of lumbar spine bone mineral content in normal women. *Calcif Tissue Int* 35:145–7.

Kruse, K. (1995). Pathophysiology of calcium metabolism in children with vitamin D-deficiency rickets. *J Pediatr* 126:736–41.

Kubota, Y., Inagaki, S., Shimada, S., Girgis, S., Zadi, M., MacIntyre, I., Tohyama, M., Kito, S. (1998). Ontogeny of the calcitonin gene-related peptide in the nervous system of rat brain stem: an immunohistochemical analysis. *Neuroscience* 26:905–26.

Kucharczyk, J. (1984a). Neuroendocrine mechanisms mediating fluid intake during the estrous cycle. *Brain Res Bull* 12:175–80.

Kucharczyk, J. (1984b). Localization of central nervous system structures mediating extracellular thirst in the female rat. *J Endocrinol* 100:183–8.

Kuga, M. (1996). A study of changes in gustatory sense during pregnancy (in Japanese). *Ribbon Jibiinkoka Gakkai Kaiho* 99:1208–17.

Kushner, R.F. (1995). Barriers to providing nutrition counseling by physicians: a survey of primary care practitioners. *Prev Med* 24:546–52.

Lackey, C.J. (1983). Pica during pregnancy. *Contemp Nutr* 8:1–2.

Lalau, J.D., Jans, I., el Esper, N., Bouillon, R., Fournier, A. (1993). Calcium metabolism, plasma parathyroid hormone, and calcitriol in transient hypertension of pregnancy. *Am J Hypertens* 6:522–7.

Landman, J.P., Hall, J. (1989). Dietary patterns and nutrition in pregnancy in Jamaica. *J Trop Pediatr* 35:185–90.

Lane, N.E., Kimmel, D.B., Nilsson, M.H., Cohen, F.E., Newton, S., Nissenson, R.A., Strewley, G.J. (1996). Bone-selective analogs of human PTH(1–34) increase bone formation in an ovariectomized rat model. *J Bone Miner Res* 11:614–25.

Lashley, K.S. (1938). An experimental analysis of instinctive behavior. *Psychol Rev* 45:445–71.

Laskey, M.A., Dibba, B., Prentice, A. (1991). Low ratios of calcium to phosphorus in the breast-milk of rural Gambian mothers. *Acta Paediatr (Oslo)* 80:250–1.

Laskey, M.A., Prentice, A., Hanratty, L.A., Jarjou, L.M., Dibba, B., Beavan, S.R., Cole, T.J. (1998). Bone changes after 3 mo of lactation: influence of calcium intake, breast-milk output, and vitamin D-receptor genotype. *Am J Clin Nutr* 67:685–92.

Laskey, M.A., Prentice, A., Jarjou, L.M., Beavan, S. (1996). Lactational changes in bone mineral of the lumbar spine are influenced by breast-milk output but not calcium intake, breast-milk calcium concentration, or vitamin-D receptor genotype. *J Bone Miner Res* 11:1815 (abstract).

Laskey, M.A., Prentice, A., Shaw, J., Zachou, T., Ceesay, S.M., Vasquez-Valesquez, L., Fraser, D.R. (1990). Breast-milk calcium concentrations during prolonged lactation in British and rural Gambian mothers. *Acta Paediatr (Oslo)* 79:507–12.

Lat, J. (1967). Self-selection of dietary components. In: *Handbook of Physiology:*

*Alimentary Canal*, C.F. Code (ed.), pp. 367–86. Washington, DC: American Physiological Society.

LeBlanc, A., Schneider, V., Spector, E., Evans, H., Rowe, R., Lane, H., Demers, L., Lipton, A. (1995). Calcium absorption, endogenous excretion, and endocrine changes during and after long-term bed rest. *Bone* 16:301S–4S.

Lee, S.J., Kanis, J.A. (1994). An association between osteoporosis and premenstrual and postmenstrual symptoms. *Bone Miner* 24:127–34.

Lehman, M.N., Powers, J.B., Winans, S.S. (1983). Stria terminalis lesions alter the temporal pattern of copulatory behaviour in the male golden hamster. *Behav Brain Res* 8:109–28.

Lehman, M.N., Winans, S.S., Powers, J.B. (1980). Medial nucleus of the amygdala mediates chemosensory control of male hamster sexual behavior. *Science* 240:557–60.

Leibowitz, S.F. (1995). Brain peptides and obesity: pharmacological treatment. *Obes Res [Suppl 4]* 3:573S–89S.

LeMagnen, J. (1983). Body energy balance and food intake: a neuroendocrine regulatory mechanism. *Physiol Rev* 63:314–86.

LeMagnen, J. (1985). *Hunger.* Cambridge University Press.

Leshem, M. (1999). Ontogeny of salt hunger in the rat. *Neurosci Biobehav Rev* 23:649–59.

Leshem, M., Boggan, B., Epstein, A.N. (1988). The ontogeny of drinking evoked by activation of brain angiotensin in the rat pup. *Dev Psychobiol* 21:63–75.

Leshem, M., Del Canho, S., Schulkin, J. (eds.) (1996). *A Possible Role for Vitamin D in Calcium Appetite.* Proceedings of Comparative Nutrition Society symposium.

Leshem, M., Del Canho, S., Schulkin, J. (1999a). Calcium hunger in the parathyroidectomized rat. *Physiol Behav* 67:555–9.

Leshem, M., Del Canho, S., Schulkin, J. (1999b). Ontogeny of calcium preference in the parathyroidectomized rat. *Dev Psychobiol* 34:293–301.

Leshem, M., Neufeld, M., Del Canho, S. (1994). Ontogeny of the ionic specificity of sodium appetite in the rat pup. *Dev Psychobiol* 27:381–94.

Leshem, M., Schulkin, J. (1998). Calcium preference during the lifespan of rats. Paper presented to the Society for the Study of Ingestive Behavior.

Leshner, A.I., Siegel, H.L. (1972). Dietary self-selection by pregnant and lactating rats. *Physiol Behav* 8:151–4.

LeVaillant, L. (1796). *Travels into the Interior Parts of Africa in the Years 1781–1785*, G.G. and J. Robinson (eds.), 2nd ed. London.

Levin, T., Schulkin, J., Leshem, M. (1996). Calcium taste in humans. *Isr Med J* 32:S41.

Levin, T., Schulkin, J., Leshem, M. Changes in intake and hedonics of minerals during pregnancy and lactation in the rat. Manuscript under review.

Levine, R.J., Hauth, J.C., Curet, L.B., Sibai, B.M., Catalano, P.M., Morris, C.D., DerSimonian, R., Esterlitz, J.R., Raymond, E.G., Bild, D.E., et al. (1997). Trial of calcium to prevent preeclampsia. *N Engl J Med* 337:69–76.

Lewis, M. (1964). Behavior resulting from calcium deprivation in parathyroidectomized rats. *J Comp Physiol Psychol* 57:348–52.

Lewis, M. (1968). Discrimination between drives for sodium chloride and calcium. *J Comp Physiol Psychol* 65:208–12.

Lichtenstein, P., Specker, B.L., Tsang, R.C., Mimouni, F., Gormley, C. (1986). Calcium-regulating hormones and minerals from birth to 18 months of age: a cross-sectional study. l. Effects of sex, race, age, season, and diet on vitamin D status. *Pediatrics* 77:883–90.

Lin, S.H, Lin, Y.F., Shieh, S.D. (1996). Milk-alkali syndrome in an aged patient with osteoporosis and fractures. *Nephron* 73:496–7.

Lindsay, R., Nieves, J., Formica, C., Hennerman, E., Woelfert, L., Shen, V., Dempster, D. (1997). Randomised controlled study of effect of parathyroid hormone on vertebral-bone mass and fracture incidence among postmenopausal women on oestrogen with osteoporosis. *Lancet* 350:550–5.

Lippuner, K., Zehnder, H.J., Cazez, J.P., Takkinen, R., Jaeger, P. (1995). Effects of PTH-related protein (PTH-P) on calcium-phosphate metabolism in nursing mothers. *Bone [Suppl 1]* 16:209S (abstract).

Lissner, L., Bengtsson, C., Hannson, T. (1991). Bone mineral content in relation to lactation history in pre- and postmenopausal women. *Calcif Tisssue Int* 48:319–25.

Liu, J., Zhu. J.K. (1998). A calcium sensor homolog required for plant salt tolerance. *Science* 180:1943–5.

Llach, F., Keshav, G., Goldblat, M.V., Lindberg, J.S., Sadler, R., Delmez, J., Arruda, J., Lau, A., Slatopolsky, E. (1998). Suppression of parathyroid hormone secretion in hemodialysis patients by a novel vitamin D analogue: 19-nor-1,25-dihydroxyvitamin $D_2$. *Am J Kidney Dis [Suppl 2]* 32:48–54.

Lobaugh, B., Joshua, I.G., Mueller, W.J. (1981). Regulation of calcium appetite in broiler chickens. *J Nutr* 111:298–306.

Lopez, J.M., Gonzalez, G., Reyes, V., Campino, C., Diaz, S. (1996). Bone turnover and density in healthy women during breastfeeding and after weaning. *Osteoporos Int* 6:153–9.

Lu, C., Ikeda, K., Deftos, L.J., Gazdar, A.F., Mangin, M., Broadus, A.E. (1989). Glucocorticoid regulation of parathyroid hormone-related peptide gene transcription in a human neuroendocrine cell line. *Mol Endocrinol* 3:2034–40.

Ludman, E.K., Kang, K.L., Lynn, L.L. (1992). Food beliefs and diets of pregnant Korean-American women. *J Am Diet Assoc* 92:1519–20.

Luine, V.N., Sonnenberg, J., Christakos, S. (1987). Vitamin D: is the brain a target? *Steroids* 49:133–53.

Lund, B., Selnes, A. (1979). Plasma 1,25-dihydroxyvitamin D levels in pregnancy and lactation. *Acta Endocrinol* 92:330–5.

Ma, L.Y., McEwen, B.S., Sakai, R.R., Schulkin, J. (1993). Glucocorticoids facilitate mineralocorticoid-induced sodium intake in the rat. *Horm Behav* 27:240–50.

McBurnie, M., Denton, D., Tarjan, E. (1988). Influence of pregnancy and lactation on Na appetite of BALC/c mice. *Am J Physiol* 255:R1020–4.

McCance, R.A. (1936). Medical problems in mineral metabolism. Experimental human salt deficiency. *Lancet* 230:823–30.

McCance, R.A. (1938). The effect of salt deficiency in man on the volume of the extracellular fluids, and on the composition of sweat, saliva, gastric juice and cerebrospinal fluid. *J Physiol* 92:208–18.

McCance, R.A., Widdowson, E.M. (1942). Mineral metabolism of healthy adults on white and brown bread dietaries. *J Physiol* 101:44–85.

McCarron, D.A. (1983). Calcium and magnesium nutrition in human hypertension. *Ann Intern Med* 98:800–5.

McCarron, D.A. (1998). Calcium supplementation and adverse outcomes of pregnancy:

a review. In: *Maternal Nutrition: New Developments and Implications.* Abstracts of an international symposium, Paris, June 11–12.

McCarron, D.A., Morris, C.D. (1985). Blood pressure response to oral calcium in persons with mild to moderate hypertension: a randomized, double-blind, placebo-controlled, crossover trial. *Ann Intern Med* 103:825–31.

McCarron, D.A., Morris, C.D., Young, E., Roullet, C., Drucke, T. (1991). Dietary calcium and blood pressure: modifying factors in specific populations. *Am J Clin Nutr* 54: 215S–19S.

McCarthy, M.M., Curran, G.H., Siegel, H.I. (1994). Evidence for the involvement of prolactin in the maternal behavior of the hamster. *Physiol Behav* 55:181–4.

McCary, L.C., DeLuca, H.F. (1996). Osteopetrotic (op/op) mice are unable to maintain serum calcium levels despite hyperabsorption of calcium. *Endocrinology* 137: 1049–56.

McCaughey, S.A., Scott, T.R. (1998). The taste of sodium. *Neurosci Biobehav Rev* 22: 663–76.

McCaughey, S.A., Tordoff, M.G. (2000). Calcium-deprived rats sham-drink $CaCl_2$ and NaCl. *Appetite* 34:305–11.

McCay, C.M., Eaton, E.M. (1947). The quality of the diet and the consumption of sugar solutions. *J Nutr* 34:351–62.

McCormick, S.D., Hasegawa, S., Hirano, T. (1992). Calcium uptake in the skin of a freshwater teleost. *Proc Natl Acad Sci USA* 89:3635–8.

McEwen, B.S. (1995). Neuroendocrine interactions. In: *Psychopharmacology: The Fourth Generation of Progress,* F.E. Bloom, D.J. Kupfer (eds.). New York: Raven Press.

McEwen, B.S. (1998). Stress, adaptation, and disease. Allostasis and allostatic load. *Ann NY Acad Sci* 840:33–44.

McEwen, B.S., Lieberburg, I., Chaptal, C., Krey, L.C. (1977). Aromatization: important for sexual differentiation of the neonatal rat brain. *Horm Behav* 9:249–63.

MacInnes, D.G., Laszlo, I., MacIntyre, I., Fink, G. (1982). Salmon calcitonin in lizard brain: a possible neuroendocrine transmitter. *Brain Res* 251:371–3.

McKane, W.R., Khosla, S., Egan, K.S., Robins, S.P., Burritt, M.F., Riggs, B.L. (1996). Role of calcium intake in modulating age-related increases in parathyroid function and bone resorption. *J Clin Endocrinol Metab* 81:1699–703.

Mackey, A.D., Picciano, M.F., Mitchell, D.C., Smiciklas-Wright, H. (1998). Self-selected diets of lactating women often fail to meet dietary recommendations. *J Am Diet Assoc* 98:297–302.

Mannstadt, M., Drueke, T.B. (1997). Recepteurs de l'hormone parathyroidienne: du clonage aux implications physiologiques, physiopathologiques et cliniques. (Parathyroid hormone receptors: from cloning to physiological, physiopathological and clinical implications) (editorial). *Nephrologie* 18(1):5–10.

Marcus, P.M., Newcomb, P.A. (1998). The association of calcium and vitamin D, and colon and rectal cancer in Wisconsin women. *Int J Epidemiol* 27:788–93.

Marcus, R. (1996). Endogenous and nutritional factors affecting bone. *Bone (Suppl)* 28:11S–13S.

Margen, S., Chu, J.Y., Kaufmann, N.A., Calloway, D.H. (1974). Studies in calcium metabolism. I. The calciuretic effect of dietary protein. *Am J Clin Nutr* 27:584–9.

Markison, S., St. John, S.J., Spector, A.C. (1995). Glossopharyngeal nerve transection does not compromise the specificity of taste-guided sodium appetite in rats. *Am J Physiol* 269:R215–21.

Markowitz, M.E., Arnaud, S., Rosen, J.F., Thorpy, M., Laximinarayan, S. (1988). Temporal interrelationships between the circadian rhythms of serum parathyroid hormone and calcium concentrations. *J Clin Endocrinol Metab* 67:1068–73.

Marlow, R.W, Tollestrup, K. (1982). Mining and exploitation of natural mineral deposits by the desert tortoise, *Gopherus agassizii*. *Animal Behavior* 32:475–8.

Martin, T.J., Grill, V. (1995). Hypercalcemia. *Clin Endocrinol* 42:535–8.

Massey, L.K., Wise, K.J. (1984). The effect of dietary caffeine on urinary excretion of calcium, magnesium, sodium, and potassium in healthy young females. *Nutr Res* 4:43–50.

Massi, M., Gentili, L., Perfumi, M., de Caro, G., Schulkin, J. (1990). Inhibition of salt appetite in the rat following injection of tachykinins into the medial amygdala. *Brain Res* 513:1–7.

Mathe, A.A., Jousisto-Hanson, J., Stenfors, C., Theodorsson, E. (1990). Effect of lithium on tachykinins, calcitonin gene-related peptide, and neuropeptide Y in rat brain. *J Neurosci Res* 26:233–7.

Matkovic, V. (1991). Calcium metabolism and calcium requirements during skeletal modeling and consolidation of bone mass. *Am J Clin Nutr* 54:245S–60S.

Matkovic, V., Fontana, D., Tominac, C., Goel, P., Chesnut, C.H., III (1990). Factors that influence peak bone mass formation: a study of calcium balance and the inheritance of bone mass in adolescent females. *Am J Clin Nutr* 52:878–88.

Matkovic, V., Heaney, R.P. (1992). Calcium balance during human growth: evidence for threshold behavior. *Am J Clin Nutr* 55:992–6.

Matkovits, T, Christakos, S. (1995). Variable in vivo regulation of rat vitamin D-dependent genes (osteopontin, Ca,Mg-adenosine triphosphatase, and 25-hydroxy-vitamin $D_3$ 24-hydroxylase): implications for differing mechanisms of regulation and involvement of multiple factors. *Endocrinology* 136:3971–82.

Matsui, H., Aou, S., Ma, J., Hori, T. (1995). Central actions of parathyroid hormone on blood calcium and hypothalamic neuronal activity in the rat. *Am J Physiol* 268: R21–7.

Mayr, E. (1982). *Systematics and the Origins of the Species*, 2nd ed. New York: Columbia University press. (Originally published 1942.)

Mayr, E. (1992). *One Long Arugument*. Cambridge, MA: Harvard University Press.

Mead, M. (1943). The problem of changing food habits. *Bulletin of the National Research Council* 108:20–31.

Melton, M.E., D'Anza, J.J., Wimbiscus, S.A., Grill, V., Martin, T.J., Kukreja, C. (1990). Parathyroid hormone-related protein and calcium homeostasis in lactating mice. *Am J Physiol* 259:E792–6.

Michell, A.R. (1975). Changes of sodium appetite during the estrous cycle of sheep. *Physiol Behav* 14:223–6.

Michell, A.R. (1976). Relationships between individual differences in salt appetite of sheep and their plasma electrolyte status. *Physiol Behav* 17:215–19.

Michell, A.R. (1986). The gut: the unobtrusive regulator of sodium balance. *Perspect Biol Med* 29:203–13.

Michell, A.R., Moss, P. (1988). Salt appetite during pregnancy in sheep. *Physiol Behav* 42:491–3.

Michelson, D., Stratakis, C., Hill, L., Reynolds, J., Galliven, E., Chrousos, G., Gold, P. (1996). Bone mineral density in women with depression. *N Engl J Med* 335:1176–81.

Mickelson, P.A., Christian, D.P. (1991). Avoidance of high-potassium diets by captive meadow voles. *J Mammol* 72:177–82.

Millelire, L., Woodside, B. (1989). Factors influencing the self-selection of calcium in lactating rats. *Physiol Behav* 46:429–34.

Miller, B.K., Latvaitis, J.A. (1992). Use of roadside salt licks by moose in northern New Hampshire. *Can Field-Natur* 106:112–17.

Miller, F.R., Sherrington, C.S. (1915). Some observations on the bucco-pharyngeal reflex deglutition in the cat. *Q J Exp Physiol* 9:147–86.

Milner, P. Zucker, I. (1965). Specific hunger for potassium in the rat. *Psychon Sci* 2:17–18.

Mitchell, D., Winter, W., Morisaki, C.M. (1977). Conditioned taste aversions accompanied by geophagia: evidence for the occurrence of "psychological" factors in the etiology of pica. *Psychosom Med* 39:401–12.

Molist, P., Rodriguez-Moldes, I., Batten, T.F., Anadon, R. (1995). Distribution of calcitonin gene-related peptide-like immunoreactivity in the brain of the small-spotted dogfish, *Scyliorhinus canicula* L. *J Comp Neurol* 352:335–50.

Monk, R.D., Bushinsky, D.A. (2000). Treatment of calcium, phosphorus and magnesium disorders. In: *Disorders of Fluid, Electrolyte and Acid–Base Balance*, H. Brady, C. Wilcox (eds.). Philadelphia: Saunders.

Mook, D.G. (1963). Oral and postingestional determinants of the intake of various solutions in rats with esophageal fistulas. *J Comp Physiol Psychol* 56(4):645–59.

Moor, C.L., Lux, B.A. (1998). Effects of lactation on sodium intake in Fischer-344 and Long-Evans rats. *Dev Psychobiol* 32:51–6.

Moore-Ede, M.C. (1986). Physiology of the circadian timing system: predictive versus reactive homeostasis. *Am J Physiol* 250:R737–52.

Moore-Ede, M.C., Burr, R.G. (1973). Circadian rhythm of urinary calcium excretion during immobilization. *Aerospace Med* 44:495–8.

Moore-Ede, M.C., Sulzman, F.M., Fuller, C.A. (1982). *The Clocks that Time Us*. Cambridge, MA: Harvard University Press.

Moos, R.H. (1968). The development of a menstrual distress questionnaire. *Psychsom Med* 30:853–67.

Morgan, C. L. (1894). *An Introduction to Comparative Psychology*. London: Walter Scott.

Morgans, C.W., El Far, O., Berntson, A., Wassle, H., Taylor, W.R. (1998). Calcium extrusion from mammalian photoreceptor terminals. *J Neurosci* 18:2467–74.

Morrison, G.R. (1971). Effects of formalin-induced Na deficiency on CaCl and KCl acceptability. *Psychon Sci* 25:167–8.

Moser, P.B., Reynolds, R.D., Acharya, S., Howard, M.P., Andon, M.B. (1988). Calcium and magnesuim dietary intakes and plasma and milk concentrations of Nepalese lactating women. *Am J Clin Nutr* 47:735–9.

Moser-Veillon, P.B., Vieira, N.E., Yergey, A.L., Nagey, D.A., Patterson, K.Y., Veillon, C. (1989). Fractional absorption and urinary excretion of calcium stable isotopes in lactating and nonlactating women. *FASEB J* 3:A645 (abstract).

Mulroney, S.E., Haramati, A. (1990). Renal adaptations to changes in dietary phosphate during development. *Am J Physiol* 258:F1650–6.

Mulroney, S.E., Woda, C.B., Halaihel, N., Louie, B., McDonnell, K., Schulkin, J., Levi, M. (in press). Regulation of type-II sodium-phosphate (NaPi-2) transporters in the brain: central control of renal NaPi-2 transporters. *Am J Physiol*.

Mundy, G.R., Guise, T.A. (1997). Hypercalcemia of malignancy. *Am J Med* 103:134–45.

Murcott, A. (1982). The cultural significance of food and eating. *Proc Nutr Soc* 41:203–10.

Musiol, I.M., Stumpf, W.E., Bidmon, H.-J., Heiss, C., Mayerhofer, A., Bartke, A. (1992). Vitamin D nuclear binding to neurons of the septal, substriatal and amygdaloid area in the Siberian hamster (*Phodopus sungorus*) brain. *Neuroscience* 48:841–8.

Myrick, A.G., Jr. (1988). Is tissue resorption and replacement in permanent teeth of mammals caused by stress-induced hypocalcemia? In: *The Biological Mechanisms of Tooth Eruption and Root Resorption*, Z. Davidovitch (ed.). Birmingham, AL: EBSCO Media.

Myrick, A.G., Jr., Stuntz, W.E., Ridgway, S.H., Odell, D.K. (1987). Hypocalcemia in spotted dolphins (*Stenella attenuata*) chased and captured by a purse seiner in the eastern tropical Pacific. In: *Abstracts of the 7th Biennial Conference on the Biology of Marine Mammals*, Miami, FL, December 5, 1987.

Nachman, M. (1963). Taste preferences for lithium chloride by adrenalectomized rats. *Am J Physiol* 205:219–21.

Nachman, M., Valentino, D. (1966). Roles of the taste and postingestional factors in the satiation of sodium appetite in rats. *Comp Physiol Psychol* 62(2):280–3.

Nakamura, K., Norgren, R. (1995). Sodium-deficient diet reduces gustatory activity in the nucleus of the solitary tract of behaving rats. *Am J Physiol* 269:R647–61.

Nakamuta, H., Itokazu, Y., Koida, M., Orlowski, R.C., Epand, R.M. (1991). Autoradiographic localization of human calcitonin sensitive binding sites in rat brain. *Jpn J Pharmacol* 56:551–5.

Namgung, R., Mimouni, F., Campaigne, B.N., Ho, M.L., Tsang, R.C. (1992). Low bone mineral content in summer-born compared with winter-born infants. *J Pediatr Gastroenterol Nutr* 15:285–8.

Namgung, R., Tsang, R.C., Specker, B.L., Sierra, R.I., Ho, M.L. (1994). Reduced serum osteocalcin and 1,25-dihydroxyvitamin D concentrations and low bone mineral content in small for gestational age infants: evidence of decreased bone formation rates. *J Pediatr* 122:269–75.

Nelson, R.J. (1995). *An Introduction to Behavioral Endocrinology*. Sunderland, MA: Sinauer Associates.

Newhall-Perry, K., Holloway, L., Osburn, L., Monroe, S.E., Heinrichs, L., Henzl, M., Marcus, R. (1995). Effects of a gonadotropin-releasing hormone agonist on the calcium-parathyroid axis and bone turnover in women with endometriosis. *Am J Obstet Gynecol* 173:824–9.

Nicolaidis, S. (1980). Hypothalamic convergence of external and internal stimulation leading to early ingestive and metabolic responses. *Brain Res Bull* [Suppl 4] 5:97–101.

Nicolaidis, S., Galaverna, O., Metzler, C.H. (1990). Exracellular dehydration during pregnancy increases salt appetite of offspring. *Am J Physiol* 258:R281–3.

Niegowska, J., Barylko-Pikleina, N. (1998). Salt taste perception in women during physiological pregnancy. *Gineko Pol* 69:168–74.

Nielsen, H.K., Laurberg, P., Brixen, K., Mosekilde, L. (1991). Relations between diurnal variations in serum osteocalcin, cortisol, parathyroid hormone, and ionized calcium in normal individuals. *Acta Endocrinol* 124:391–8.

Nieves, J.W., Golden, A.L., Siris, E., Kelsey, J.L., Lindsay, R. (1995). Teenage and current calcium intake are related to bone mineral density of the hip and forearm in women aged 30–39 years. *Am J Epidemiol* 141:342–51.

Nieves, J.W., Komar, L., Cosman, F., Lindsay, R. (1998). Calcium potentiates the effect of estrogen and calcitonin on bone mass: review and analysis. *Am J Clin Nutr* 67:18–24.

Nishijo, H., Norgren, R. (1997). Parabrachial neural coding of taste stimuli in awake rats. *J Neurophysiol* 78:2254–68.

Nitabach, M., Schulkin, J., Epstein, A.N. (1989). The medial amygdala is part of a mineralocorticoid-sensitive circuit controlling NaCl intake in the rat. *Behav Brain Res* 15:197–204.

Njwe, R.M., Kom, J. (1988). Survey of the mineral status of pastures and small ruminants in the west region of Cameroon. *Tropicultura* 6:150–2.

Nordin, B.E.C. (1988). *Calcium in Human Biology, Human Nutrition Reviews.* London: Springer-Verlag.

Nordin, B.E.C., Marshall, D.H. (1988). Dietary requirements for calcium. In: *Calcium in Human Biology*, B.E.C. Nordin (ed.), pp. 447–64. London: Springer-Verlag.

Norgren, R. (1976). Taste pathways to hypothalamus and amygdala. *J Comp Neurol* 166:17–30.

Norgren, R. (1984). Central neural mechanisms of taste. In: *Handbook of Physiology and the Nervous System. Vol. 3: Sensory Processes*, pt. 2.1, J.M. Brookhart, V.B. Mountcastle (eds.), pp. 1087–128. Bethesda, MD: American Physiological Society.

Norgren, R. (1985). The sense of taste and the study of ingestion. In: *Taste, Olfaction, and the CNS*, D.W. Pfaff (ed.). New York: Rockefeller University Press.

Norgren, R. (1995). Gustatory system. In: *The Rat Nervous System.* Orlando: Academic Press.

Norgren, R., Schulkin, J., Grigson, P.S. (1999). Parabrachial nucleus lesions increase NaCl intake in calcium deprived rats. *Soc Neurosci Abstr.*

Norgren, R., Wolf, G. (1975). Projections of thalamic gustatory and lingual areas in the rat. *Brain Res* 92:123–9.

Norman, A.W. (1995). Transcaltachia (the rapid hormonal stimulation of intestinal calcium transport): a component of adaptation to calcium needs and calcium availability. *Am Zool* 35:483–9.

Norris, C.M., Halpain, S., Foster, T.C. (1998). Reversal of age-related alterations in synaptic plasticity by blockade of L-type $Ca^{2+}$ channels. *J Neurosci* 18:3171–9.

Nowlis, G.H. (1977). From reflex to representation: taste-elicited tongue movements in the human newborn. Chapter 13 in *Taste and Development*, J.M. Weiffenbach (ed.). Bethesda, MD: U.S. DHEW.

Nutley, M.T., Parimi, S.A., Harvey, S. (1995). Sequence analysis of hypothalamic parathyroid hormone messenger ribonucleic acid. *Endocrinology* 136:5600–7.

Oftedal, O.T. (1984). Milk composition, milk yield, and energy output at peak lactation: a comparative review. *Symp Zool Soc London* 51:33–85.

Oftedal, O.T. (1991). The nutritional consequences of foraging in primates: the relationship of nutrient intakes to nutrient requirements. *Philos Trans R Soc Lond Biol Sci* 334:161–70.

Oftedal, O.T., Chen, T.C., Schulkin, J. (1997). Preliminary observations on the relationship of calcium ingestion to vitamin D status in the green iguana (*Iguana iguana*). *Zool Biol* 16:201–7.

Oguro, C., Sasayama, Y. (1978). Function of the parathyroid gland in serum calcium regulation in the newt *Tylototriton andersoni* Boulenger. *Gen Comp Endocrinol* 35:10–15.

Ohkura, T., Isse, K., Akazawa, K., Hamamoto, M., Yaoi, Y., Hagino, N. (1994). Evaluation of estrogen treatment in female patients with dementia of the Alzheimer type. *Endocr J* 41:361–71.

Okawara, Y., Kobayashi, H. (1986). Biologically active peptides and water intake. I. Water intake and excretion induced by parathyroid hormone in the Japanese quail *Coturnix coturnix japonica*. *Zool Sci* 3:825–30.

Okiyama, A., Torii, K., Tordoff, M.G. (1996). Increased NaCl preference of rats fed low-protein diet. *Am J Physiol* 39:R1189–96.

Ono, T., Inokuchi, K., Ogural, A., Ikawa, Y., Kudo, Y., Kawashima, S. (1997). Activity-dependent expression of parathyroid hormone-related protein (PTHrP) in rat cerebellar granule neurons. *J Biol Chem* 272:14404–11.

Ooms, M.E., Roos, J.C., Bezemer, P.D., Van de Vijgh, W.J.F., Bouter, L.M., Lips, P. (1995). Prevention of bone loss by vitamin D supplementation in elderly women: a randomized double-blind trial. *JAMA* 80:1052–8.

Orwoll, E.S. (1982). The milk-alkali syndrome: current concepts. *Ann Intern Med* 97:242–8.

Osborne, T.B., Mendel, L.B. (1918a). The choice between adequate and inadequate diets as made by rats. *J Biol Chem* 35:19–27.

Osborne, T.B., Mendel, L.B. (1918b). The inorganic elements in nutrition. *Am J Physiol* 205:30–48.

Packard, M.J. (1992). Use of slow-release pellets to administer calcitriol to avian embryos; effects on plasma calcium, magnesium, and phosphorus. *Gen Comp Endocrinol* 85:8–16.

Packard, M.J. (1994a). Patterns of mobilization and deposition of calcium in embryos of oviparous, amniotic vertebrates. *J Zool* 40:481–92.

Packard, M.J. (1994b). Mobilization of shell calcium by the chick chorioallantoic membrane in vitro. *J Exp Biol* 190:141–53.

Packard, M.J., Clark, N.B. (1996). Aspects of calcium regulation in embryonic lepidosaurians and chelonians and a review of calcium regulation in embryonic archosaurians. *Physiol Zool* 69:435–66.

Pang, P.K., Harvey, S., Fraser, R., Kaneko, T. (1988). Parathyroid hormone-like immunoreactivity in brains of tetrapod vertebrates. *Am J Physiol* 255:R635–42.

Parent, M.E., Krondl, M., Chow, R.K. (1993). Reconstruction of past calcium intake patterns during adulthood. *J Am Diet Assoc* 93:649–52.

Parfitt, A.M. (1977). Metacarpal cortical dimensions in hypoparathyroidism, primary hyperparathyroidism and chronic renal failure. *Calcif Tissue Res* 22:329–31.

Parfitt, A.M., Chir, B., Gallagher, J.C., Heaney, R.P., Johnston, C.C., Neer, R., Whedon, G.D. (1982). Vitamin D and bone health in the elderly. *Am J Clin Nutr* 36:1014–31.

Parfitt, A.M., Higgins, B.A., Nassim, J.R., Collins, J.A., Hilb, A. (1964). Metabolic studies in patients with hypercalciuria. *Clin Sci* 27:463–82.

Parker, D., Emmett, P.M., Heaton, K.W. (1992). Final year medical students' knowledge of practical nutrition. *J R Soc Med* 85:338.

Parry-Jones, B. (1992). Pagophagia, or compulsive ice consumption: a historical perspective. *Psychol Med* 22:561–71.

Paul, A.A., Bates, C.J., Prentice, A., Day, K.C., Tsuchiya, H. (1998). Zinc and phytate intake of rural Gambian infants: contributions from breastmilk and weaning foods. *Int J Food Sci Nutr* 49:141–55.

Paul, A.A., Southgate, D.A.T. (1978). In: *The Composition of Foods*, 4th ed., R.A. McCance, E.M. Widdowson (eds.). London: HMSO.

Paulus, R.A., Eng, R., Schulkin, J. (1984). Preoperative latent place learning preserves salt appetite following damage to the central gustatory system. *Behav Neurosci* 98:146–51.

Pavlov, I.P. (1927). *Conditioned Reflexes*. New York: Dover.

Pavlovitch, H., Clemens, T.L., Laouari, D., O'Riordan, J.L., Balsan, S. (1980). Lack of effect of ovariectomy on the metabolism of vitamin D and intestinal calcium-binding protein in female rats. *J Endocrinol* 86:419–24.

Paxinos, G., Watson, C. (1982). *The Rat Brain in Sterotaxic Coordinates*. New York: Academic Press.

Peacock, M. (1991). Calcium absorption efficiency and calcium requirements in children and adolescents. *Am J Clin Nutr* 54:261S–5S.

Pearce, S.H. (1999). Extracellular "calcistat" in health and disease. *Lancet* 353:83–4.

Pearce, S.H., Brown, E.M. (1996). Calcium-sensing receptor mutations: insights into a structurally and functionally novel receptor (editorial). *J Clin Endocrinol Metab* 81:1309–11.

Pedersen, E.B., Johannesen, P., Kristensen, S., Rasmussen, A.B., Emmertsen, K., Moller, J., Lauritsen, J.G., Wohlert, M. (1984). Calcium parathyroid hormone and calcitonin in normal pregnancy and preeclampsia. *Gynecol Obstet Invest* 18:156–64.

Pelchat, M.L., Rozin, P. (1982). The special role of nausea in the acquisition of food dislikes in humans. *Appetite* 3:341–51.

Penland, J.G., Johnson, P.E. (1993). Dietary calcium and manganese effects on menstrual cycle symptoms. *Am J Obstet Gynecol* 168:1417–23.

Pennington, J.A., Wilson, D.B. (1990). Daily intakes of nine nutritional elements: analyzed vs. calculated values. *Am J Diet Assoc* 90(3):375–82.

Perry, S.F., Gross, C.G., Fenwick, J.C. (1992). Interrelationship between gill chloride cell morphology and calcium uptake in freshwater teleosts. *Physiol Biochem* 10:327–37.

Petersen, O.H., Petersen, C.C., Kasai, H. (1994). Calcium and hormone action. *Annu Rev Physiol* 56:297–319.

Petraglia, F., Volpe, A., Genazzani, A.R., Rivier, J., Sawchenko, P.E., Vale, W. (1990). Neuroendocrinology of the human placenta. *Front Neuroendocrinol* 11:6–37.

Pettifor, J.M., Ross, P., Want, J., Moodley, G., Couper-Smith, J. (1978). Rickets in children of rural origin in South Africa: is low dietary calcium a factor? *J Pediatr* 92:320–4.

Pfaff, D.W. (1980). *Estrogens and Brain Function: Neural Analysis of a Hormone-controlled Mammalian Reproductive Behavior*. Berlin: Springer-Verlag.

Pfaff, D.W., Dellovade, T., Zhu, Y.S. (1997). Interactions among genes for transcription factors in hypothalamic neurons: implications for reproductive behaviors. *Mol Psychiatry* 2:448–50.

Pfaffmann, C. (1952). Taste preference and aversion following lingual denervation. *J Comp Physiol Psychol* 45:393–400.

Pfaffmann, C. (1960). The pleasures of sensation. *Psychol Rev* 67:253–68.

Pfaffmann, C. (1965). De gustibus. *Am Psychol* 20:21–31.

Pfaffmann, C. (1967). The sense of taste. In: *Handbook of Physiology. Section 6. Alimentary Canal Vol 1. Food and Water Intake*, C.F. Code (ed.). Washington, DC: American Physiological Society.

Pfaffmann, C. (1982). Taste: a model of incentive motivation. In: *The Physiological Mechanisms of Motivation*, D.W. Pfaff (ed.), pp. 61–7. New York: Springer-Verlag.

**181**

Pfaffmann, C., Norgren, R., Grill, H.J. (1977). Sensory affect and motivation. *Ann NY Acad Sci* 290:18–34.

Phillips, R.D. (1993). Starchy legumes in human nutrition, health, and culture. *Plant Foods Hum Nutr* 44:195–211.

Phoenix, C.H., Goy, R.W., Gerall, A.A., Young, W.C. (1959). Organizing action of prenatally administered testosterone propionate on the tissues mediating mating behavior in the female guinea pig. *Endocrinology* 65:369–89.

Pietrowsky, R., Preuss, S., Born, J., Pauschinger, P., Fehm, H.L. (1989). Effects of cholecystokinin and calcitonin on evoked brain potentials and satiety in man. *Physiol Behav* 46:513–19.

Pietschmann, P., Woloszczuk, W., Pietschmann, H. (1990). Increased serum osteocalcin levels in elderly females with vitamin D deficiency. *Exp Clin Endocrinol* 95:275–8.

Pike, R.L., Yao, C. (1971). Increased sodium chloride appetite during pregnancy in the rat. *J Nutr* 101:169–76.

Pinker, S. (1994). *The Language Instinct*. New York: Morrow.

Pinker, S. (1997). *How the Mind Works*. New York: Norton.

Pitkin, R.M. (1975). Calcium metabolism in pregnancy: a review. *Am J Obstet Gynecol* 121:724–37.

Pitkin, R.M. (1983). Endocrine regulation of calcium homeostasis during pregnancy. *Clin Perinatol* 10:575–92.

Planells, E., Llopis, J., Peran, F., Aranda, P. (1995). Changes in tissue calcium and phosphorus content and plasma concentrations of parathyroid hormone and calcitonin after long-term magnesium deficiency in rats. *J Am Coll Nutr* 14:292–8.

Plouffe, L., Schulkin, J. (1998). The clinical relevance of estrogen in cognition, memory and mood. *J Soc Obstet Gynaecol Can* 201:929–41.

Pochet, R., Parmentier, M., Lawson, D.E., Pasteels, J.L. (1985). Rat brain synthesizes two vitamin D-dependent calcium-binding proteins. *Brain Res* 345:251–6.

Polatti, F., Capuzzo, E., Viazzo, F., Colleoni, R., Klersy, C. (1999). Bone mineral changes during and after lactation. *Obstet Gynecol* 94:52–6.

Polatti, F., Perotti, F., Angelini, G.P., Vassellatti, D., Rapisardi, I. (1993). Effects of salmon calcitonin suppositories in the prevention of bone loss in oophorectomized women. *Maturitas* 18:73–6.

Pompei, O., Tayebaty, S.J., DeCaro, G., Schulkin, J., Massi, M. (1992). Bed nucleus of the stria terminalis: site for the antinatriorexegenic action of tachykinins in the rat. *Pharmacol Biochem Behav* 40:977–81.

Pope, J.F., Skinner, J.D., Carruth, B.R. (1992). Cravings and aversions of pregnant adolescents. *J Am Diet Assoc* 92:1479–82.

Poppitt, S.D., Prentice, A.M., Jequier, E., Schutz, Y., Whitehead, R.G. (1993). Evidence of energy sparing in Gambian women during pregnancy: a longitudinal study using whole-body calorimetry. *Am J Clin Nutr* 57:353–64.

Potter, J.D. (1996). Nutrition and colorectal cancer. *Cancer Causes Control* 7:127–46.

Power, M.L., Heaney, R.P., Kalkwarf, H.J., Pitkin, R.M., Repke, J.T., Tsang, R.C., Schulkin, J. (1999). The role of calcium in health and disease. *Am J Obstet Gynecol* 181:1560–9.

Power, M.L., Holzman, G.B., Schulkin, J. (1999). Calcium nutrition and the practicing obstetrician-gynecologist: effects of age, gender and attitude. *Obstet Gynecol* 94:421–6.

Power, M.L., Oftedal, O.T., Savage, A., Blumer, E.S., Soto, L.H., Chen, T.C., Holick, M.F. (1997). Assessing vitamin D status of callitrichids: baseline data from wild cotton-top tamarins (*Saguinus oedipus*) in Colombia. *Zool Biol* 16:39–46.

Power, M.L., Tardif, S.D., Layne, D.G., Schulkin, J. (1999). Ingestion of calcium solutions by common marmosets (*Callithrix jacchus*). *Am J Primatol* 47:255–61.

Powley, T.L. (1977). The ventromedial hypothalamic syndrome, satiety and a cephalic phase hypothesis. *Psychol Rev* 84:89–126.

Prentice, A.M. (1991). Can maternal dietary supplements help in preventing infant malnutrition? *Acta Paediatr (Oslo) [Suppl]* 324:67–77.

Prentice, A. (1994a). Maternal calcium requirements during pregnancy and lactation. *Am J Clin Nutr [Suppl]* 59:477S–83S.

Prentice, A. (1994b). Calcium intakes and bone densities of lactating women and breast-fed infants in Gambia. In: *Nutrient Regulation during Pregnancy, Lactation, and Infant Growth*, L. Allen, J. King, B. Lonnerdal (eds.). New York: Plenum Press.

Prentice, A.M. (1998). Manipulation of dietary fat and energy density and subsequent effects on substrate flux and food intake. *Am J Clin Nutr* 67:535S–41S.

Prentice, A., Barclay, D.V. (1991). Breast-milk calcium and phosphorus concentrations of mothers in rural Zaire. *Eur J Clin Nutr* 45:611–17.

Prentice, A.M., Goldberg, G.R., Poppitt, S.D. (1996a). Reproductive stress in under-nourished and well-nourished women. *Bibliotheca Nutritio et Dieta* 53:1–10.

Prentice, A., Jarjou, L.M., Cole, T.J., Stirling, D.M., Dibba, B., Fairweather-Tait, S. (1995). Calcium requirements of lactating Gambian mothers: effects of a calcium supplement on breast-milk calcium concentration, maternal bone mineral content, and urinary calcium excretion. *Am J Clin Nutr* 62:58–67.

Prentice, A., Jarjou, L.M., Sterling, D.M., Buffenstein, R., Fairweather-Tait, S. (1998). Biochemical markers of calcium and bone metabolism during 18 months of lactation in Gambian women accustomed to a low calcium intake and in those consuming calcium supplement. *J Clin Endocrinol Metab* 83:1059–66.

Prentice, A., Laskey, M.A., Shaw, J., Cole, T.J., Fraser, D.R. (1990). Bone mineral content of Gambian and British children aged 0–36 months. *Bone Miner* 10:211–24.

Prentice, A., Laskey, M.A., Shaw, J., Hudson, G.J., Day, K.C., Jarjou, L.M., Dibba, B., Paul, A.A. (1993). The calcium and phosphorus intakes of rural Gambian women during pregnancy and lactation. *Br J Nutr* 69:885–96.

Prentice, A.M., Roberts, S.B., Paul, A.A., Watkinson, M., Watkinson, A.A., Whitehead, R.G. (1983). Dietary supplementation of lactating Gambian women. I. Effect on breast-milk volume and quality. *Hum Nutr Clin Nutr* 37:53–64.

Prentice, A.M., Spaaij, C.J., Goldberg, G.R., Poppitt, S.D., van Raaij, J.M., Totton, M., Swann, D., Black, A.E. (1996b). Energy requirements of pregnant and lactating women. *Eur J Clin Nutr* 50:S82–111.

Prentice, A., Yan, L., Jarjou, L.M.A., Dibba, B., Laskey, M.A., Stirling, D.M., Fairweather-Tait, S. (1997). Vitamin D status does not influence the breast-milk calcium concentration of lactating mothers accustomed to a low calcium intake. *Acta Paediatr (Oslo)* 86:1006–8.

Prince, R., Devine, A., Dick, I., Criddle, A., Kerr, D., Kent, N., Price, R., Randell, A. (1995). The effects of calcium supplementation (milk powder or tablets) and exercise on bone density in postmenopausal women. *J Bone Miner Res* 10:1068–75.

Puchacz, E., Stumpf, W.E., Stachowiak, E.K., Stachowiak, M.K. (1996). Vitamin D increases expression of the tyrosine hydroxylase gene in adrenal medullary cells. *Brain Res Mol Brain Res* 36:193–6.

Pun, K.K., Chan, L.W.L., Chung, V., Wong, F.H.W. (1991). Calcium content of common food items in Chinese diet. *Calcif Tissue Int* 48:153–6.

Purwar, M., Kulkarni, H., Motghare, V., Dhole, S. (1996). Calcium supplementation and prevention of pregnancy induced hypertension. *J Obstet Gynaecol Res* 22:425–30.

Putney, J.W., Jr., Bird, G. (1993). The inositol phosphate calcium signalling system in non-excitable cells. *Endocr Rev* 14:610–31.

Puzas, J.E. (1996). Osteoblast cell biology – lineage and functions. In: *Primer on the Metabolic Bone Diseases and Disorders of Mineral Metabolism*, 3rd ed., M.J. Favus (ed.). New York: Lippincott-Raven.

Quinn, S.J., Ye, C.P., Diaz, R., Kifor, O., Bai, M., Vassilev, P., Brown, E. (1997). The $Ca^{2+}$-sensing receptor: a target for polyamines. *Am J Physiol* 273:C1315–23.

Rabe, E.F., Corbit, J.D. (1973). Postingestional control of sodium chloride solution drinking in the rat. *J Comp Physiol Psychol* 84:268–74.

Ralston, S.H. (1994). Analysis of gene expression in human bone biopsies by polymerase chain reaction: evidence for enhanced cytokine expression in postmenopausal osteoporosis. *J Bone Miner Res* 9:883–90.

Ramsdale, S.J., Bassey, E.J., Pye, D.J. (1994). Dietary calcium intake relates to bone mineral density in premenopausal women. *Br J Nutr* 71:77–84.

Rasmussen, P. (1977). Calcium deficiency, pregnancy and lactation in rats: histological microradiographical and fluorescence microscopical observations on mandibular bone. *J Periodont Res* 12:491–9.

Raynaud, A., Cohen, R., Modigliani, E. (1994). Le peptide alternatit du gêne de la calcitonine [Calcitonin gene-related peptide (CGRP)]. *Presse Médicale* 23:171–5.

Reber, P.M., Heath, H., III (1995). Hypocalcemic emergencies. *Med Clin North Am* 79: 93–106.

Rebut-Bonneton, C., Demignon, J., Amor, B., Miravet, L. (1983). Effect of calcitonin in pregnant rats on bone resorption in fetuses. *J Endocrinol* 9:347–53.

Recker, R.R., Davies, K.-M., Dowd, R.M., Heaney, R.P. (1999). The effect of low-dose continuous estrogen and progesterone therapy with calcium and vitamin D on bone in elderly women. A randomized, controlled trial. *Ann Intern Med* 130:897–904.

Recker, R.R., Hinders, S., Davies, K.M., Heaney, R.P., Stegman, M.R., Lappe, J.M., Kimmel, D.B. (1996). Correcting calcium nutritional deficiency prevents spine fractures in elderly women. *J Bone Miner Res* 11:1961–6.

Reed, D.B., Meeks, P.M., Nguyen, L., Cross, E.W., Garrison, M.E.B. (1998). Assessment of nutrition education needs related to increasing dietary calcium intake in low-income Vietnamese mothers using focus group discussions. *J Nutr Educ* 30:155–63.

Reed, J.A., Anderson, J.J.B., Tylavsky, F.A., Gallagher, P.N., Jr. (1994). Comparative changes in radial-bone density of elderly female lactoovovegetarians and omnivores. *Am J Clin Nutr* 59:1197S–202S.

Reeve, J. (1980). Calcium metabolism. In: *Clinical Physiology in Obstetrics*, F. Hytten, G. Chamberlain (eds.), pp. 257–69. Oxford: Blackwell.

Reichel, H., Koeffler, H.P., Norman, A.W. (1989). The role of the vitamin D endocrine system in health and disease. *N Engl J Med* 320:980–91.

Reid, I.R. (1996). Therapy of osteoporosis: calcium, vitamin D, and exercise. *Am J Med Sci* 312:278–86.

Reid, I.R., Ames, R.A., Evans, M.C., Gamble, G.D., Sharpe, S.J. (1995). Long-term effects of calcium supplementation on bone loss and fractures in postmenopausal women: a randomized controlled trial. *Am J Med* 98:331–5.

Reilly, J.J., Maki, R., Nardozzi, J., Schulkin, J. (1994). The effects of lesions of the bed nucleus of the stria terminalis on sodium appetite. *Acta Neurobiol Exp* 54:253–7.

Reilly, J.J., Nardozzi, J., Schulkin, J. (1995). The ingestion of calcium in multiparous and virgin female rats. *Brain Res Bull* 37:301–3.

Reilly, J.J., Schulkin, J. (1993). Hormonal control of calcium ingestion: the effects of neonatal manipulations of the gonadal steroids both during and after critical stages in development on the calcium ingestion of male and female rats. *Psychobiology* 21:50–4.

Remington, R.E. (1936). The social origins of dietary ingestion. *Scientific Monthly* 12:253–68.

Repke, J.T. (1994a). Calcium and vitamin D. *Clin Obstet Gynecol* 37:550–7.

Repke, J.T. (1994b). Calcium homeostasis in pregnancy. *Clin Obstet Gynecol* 37:59–65.

Repke, J.T., Robinson, J.N. (1998). The prevention and management of pre-eclampsia and eclampsia. *Int J Gynaecol Obstet* 62:1–9.

Repke, J.T., Villar, J. (1991). Pregnancy-induced hypertension and low birth weight: the role of calcium. *Am J Clin Nutr* 54:237S–41S.

Rescorla, R.A. (1981). Simultaneous associations. In: *Predictability, Correlation and Contingency*, P. Harzem, M.D. Zeiler (eds.). New York: Wiley.

Rescorla, R.A., Freberg, L. (1978). The extinction of within-compound flavor associations. *Learn Motiv* 9:411–27.

Retallack, R.W., Jeffries, M., Kent, G.N., Hitchcock, N.E., Gutteridge, D.H., Smith, M. (1977). Physiological hyperparathyroidism in human lactation. *Calcif Tissue Res* [*Suppl*]22:142–6.

Ricci, A.J., Wu, Y.C., Fettiplace, R. (1998). The endogenous calcium buffer and the time course of transducer adaptation in auditory hair cells. *J Neurosci* 18:8261–77.

Richter, C.P. (1936). Increased salt appetite in adrenalectomized rats. *Am J Physiol* 115:155–61.

Richter, C.P. (1943). Total self-regulatory functions in animals and human beings. *Harvey Lect* 38:63–103.

Richter, C.P. (1953). Experimentally produced reactions to food poisoning in wild and domesticated rats. *Ann NY Acad Sci* 56:225–39.

Richter, C.P. (1955). Self-regulatory functions during gestation and lactation. In: *Gestation*, C.A. Villee (ed.). New York: Josiah Macy, Jr., Foundation.

Richter, C.P. (1956). Salt appetite of mammals: its dependence on instinct and metabolism. In: *L'Instinct dans le Comportement des Animaux et de l'Homme*, pp. 577–629. Paris: Masson.

Richter, C.P. (1965). *Biological Clocks in Medicine and Psychiatry*. Springfield, IL: Charles C. Thomas.

Richter, C.P., Barelare, B., Jr. (1938). Nutritional requirements of pregnant and lactating rats studied by the self-selection method. *Endocrinology* 23:15–24.

Richter, C.P., Barelare, B., Jr. (1939). Further observations on the carbohydrate, fat, and protein appetite of vitamin B deficient rats. *Am J Physiol* 127:199–210.

Richter, C.P., Birmingham, J.R. (1941). Calcium appetite of parathyroidectomized rats used to bioassay substances which affect blood calcium. *Endocrinology* 29:657–66.

Richter, C.P., Eckert, J.F. (1937). Increased calcium appetite of parathyroidectomized rats. *Endocrinology* 21:50–4.

Richter, C.P., Eckert, J.F. (1938a). Mineral appetite of parathyroidectomized rats. *Am J Med Sci* 198:9–16.

Richter, C.P., Eckert, J.F. (1938b). Mineral metabolism of adrenalectomized rats studied by the appetite method. *Endocrinology* 22:214–24.

Richter, C.P., Helfrick, S. (1943). Decreased phosphorus appetite of parathyroidectomized rats. *Endocrinology* 33:349–52.

Richter, C.P., Honeyman, W.M., Hunter, H. (1940). Behaviour and mood cycles apparently related to parathyroid deficiency. *J Neurol Neurosurg Psychiatry* 3:19–25.

Richter, C.P., Rice, K.K. (1943). Effects produced by vitamin D on energy, appetite, and oestrous cycles of rats kept on an exclusive diet of yellow corn. *Am J Physiol* 139:693–9.

Rico, H., Hernandez, E.R., Gomez-Castresana, F. (1992). Salmon calcitonin reduces vertebral fracture rate in postmenopausal crush fracture syndrome. *Bone Miner* 16:131–8.

Riggs, B.L., Melton, L.J., III (1995). The worldwide problem of osteoporosis: insights afforded by epidemiology. *Bone* 17:505S–11S (abstract).

Riggs, B.L., O'Fallon, W.M., Muse, J., O'Conner, M.K., Melton, L.J., III (1996). Long-term effects of calcium supplementation on serum PTH, bone turnover, and bone loss in elderly women. *J Bone Miner Res* 11:S118.

Riis, B., Thomsen, K., Christiansen, C. (1987). Does calcium supplementation prevent postmenopausal bone loss? *N Engl J Med* 316:173–7.

Ritchie, L.D., Fung, E.B., Halloran, B.P., Turnlund, J.R., Van Loan, M.D., Cann, C.E., King, J.C. (1998). A longitudinal study of calcium homeostasis during human pregnancy and lactation and after resumption of menses. *Am J Clin Nutr* 67:693–701.

Robertson, W.G. (1988). Chemistry and biochemistry of calcium. In: *Calcium in Human Biology*, B.E.C. Nordin (ed.). New York: Springer-Verlag.

Rodgers, W.L. (1967). Specificity of specific hungers. *J Comp Physiol Psychol* 64:49–58.

Rodgers, W.L., Rozin, P. (1966). Novel food preferences in thiamine deficient rats. *J Comp Physiol Psychol* 61:194–9.

Rogers, K.V., Dunn, C.K., Hebert, S.C., Brown, E.M. (1997). Localization of calcium receptor mRNA in the adult rat central nervous system by in situ hybridization. *Brain Res* 744:47–56.

Rogers, P.J., Edwards, S., Green, M.W., Jas, P. (1992). Nutritional influences on mood and cognitive performance; the menstrual cycle, caffeine and dieting. *Proc Nutr Soc* 51:343–51.

Roitman, M.F., Schafe, G.E., Thiele, T.E., Bernstein, I.L. (1997). Dopamine and sodium appetite: antagonists suppress sham drinking of NaCl solutions in the rat. *Behav Neurosci* 111:606–11.

Rolls, B.J., Rools, E.T., Rave, E.A. (1982). The influence of variety on human food selection and intake. In: *The Psychobiology of Human Food Selection*, L.M. Barker (ed.), pp. 101–22. Westport, CT: AVI.

Roper, S.D. (1992). The microphysiology of peripheral taste organs. *J Neurosci* 12:1127–34.

Roselle, H.A. (1970). Association of laundry starch and clay ingestion with anemia in New York City. *Arch Intern Med* 125:57–61.

Rosenblatt, J.S., Lehrman, D.S. (1963). Maternal behavior of the laboratory rat. In: *Maternal Behavior in Mammals*, H.L. Rheingold (ed.). New York: Wiley.

Rosenwasser, A.M., Adler, A.N. (1986). Structure and function in circadian timing systems. *Neurosci Biobehav Rev* 10:431–48.

Roth, J., Baetens, D., Norman, A.W., Garcia-Segura, L.M. (1981). Specific neurons in chick central nervous system stain with an antibody against chick intestinal vitamin D-dependent calcium-binding protein. *Brain Res* 222:452–7.

Rowland, N.E., Bellush, L.L., Fregly, M.J. (1985). Nycthemeral rhythms and sodium chloride appetite in rats. *Am J Physiol* 249:R375–8.

Rozin, P. (1967a). Specific aversions as a component of specific hungers. *J Comp Physiol Psychol* 64:237–42.

Rozin, P. (1967b). Thiamine specific hunger. Chapter 30 in *Handbook of Physiology – Alimentary Canal*, C.F. Code (ed.), pp. 411–31. Washington, DC: American Physiological Society.

Rozin, P. (1968). Are carbohydrate and protein intakes separately regulated? *J Comp Physiol Psychol* 65:23–9.

Rozin, P. (1976a). The selection of foods by rats, humans, and other animals. In: *Advances in the Study of Behavior*, vol. 6, J.S. Rosenblatt, R.A. Hinde, E. Shaw, C. Beer (eds.). New York: Academic Press.

Rozin, P. (1976b). The evolution of intelligence: access to the cognitive unconscious. In: *Progress in Physiological Psychology*, J. Sprague, A.N. Epstein (eds.). New York: Academic Press.

Rozin, P. (1976c). *Curt Richter: The Compleat Psychobiologist*, E.M. Blass (ed.). Baltimore: York Press.

Rozin, P., Fallon, A.E. (1987). A perspective on disgust. *Psychol Rev* 94:23–41.

Rozin, P., Levine, E., Stoess, C. (1991). Chocolate craving and liking. *Appetite* 17:199–212.

Rozin, P., Schulkin, J. (1990). Food selection. In: *Handbook of Behavioral Neurobiology*, E.M. Stricker (ed.). New York: Plenum Press.

Rudnicki, M., Thode, J., Jorgensen, T., Heitmann, B.L., Sorensen, O.H. (1993). Effects of age, sex, season and diet on serum ionized calcium, parathyroid hormone and vitamin D in a random population. *J Intern Med* 234:195–200.

St. John, S.J., Spector, A.C. (1998). Behavioral discrimination between quinine and KCl is dependent on input from the seventh cranial nerve: implications for the functional roles of the gustatory nerves in rats. *J Neurosci* 18:4353–62.

Sakai, R.R., Nicolaidis, S., Epstein, A.N. (1989). Salt appetite is suppressed by interference with angiotensin II and aldosterone. *Am J Physiol* 251:R472–8.

Samuels, M.H., Veldhuis, J.D., Kramer, P., Urban, R.J., Bauer, R., Mundy, G.R. (1997). Episodic secretion of parathyroid hormone in postmenopausal women: assessment by deconvolution analysis and approximate entropy. *J Bone Miner Res* 12:616–23.

Sandberg, A.S., Larsen, T., Sandstrom, B. (1993). High dietary calcium level decreases colonic phytate degradation in pigs fed a rapeseed diet. *J Nutr* 123:559–66.

Sapolsky, R.M. (1992). *Stress, the Aging Brain, and the Mechanisms of Neuron Death.* Cambridge, MA: MIT Press.

Saria, A., Bernatzky, G., Humpel, C., Haring, C., Skofitsch, G., Panksepp, J. (1992). Calcitonin gene-related peptide in the brain. *Ann NY Acad Sci* 657:164–9.

Sayegh, R., Schiff, I., Wurtman, J., Spiers, P., McDermott, J., Wurtman, R. (1995). The effect of a carbohydrate-rich beverage on mood, appetite, and cognitive function in women with premenstrual syndrome. *Obstet Gynecol* 80:520–8.

Scalera, G., Spector, A.C., Norgren, R. (1995). Excitotoxic lesions of the parabrachial nuclei prevent conditioned taste aversions and sodium appetite in rats. *Behav Neurosci* 109:997–1008.

Scalafani, A. (1987). Carbohydrate, taste, appetite and obestity: an overview. *Neurosci Biobehav Rev* 11:131–53.

Schaafsma, G. (1988). Calcium in extracellular fluid: homeostasis. Chapter 10 in *Calcium in Human Biology*, B.E.C. Nordin (ed.). New York: Springer-Verlag.

Schaller, G.B. (1963). *The Mountain Gorilla*. University of Chicago Press.

Scheidler, M.G., Verbalis, J.G., Stricker, E.M. (1994). Inhibitory effects of estrogen on stimulated salt appetite in rats. *Behav Neurosci* 208:141–50.

Schiff, E., Friedman, S.A., Sibai, B.M., Kao, L., Schifter, S. (1995). Plasma and placental calcitonin gene-related peptide in pregnancies complicated by severe preeclampsia. *Am J Obstet Gynecol* 173:1405–9.

Schiffman, S.S., McElroy, A.E., Erickson, R.P. (1980). The range of taste quality of sodium salts. *Physiol Behav* 24:217–24.

Schmidt-Nielsen, B. (1997). *Animal Physiology: Adaptation and Environment*, 5th ed. Cambridge University Press. (Originally published 1975.)

Schmidt-Nielsen, B., Renfro, J.L. (1975). Kidney function of the American eel *Anguilla rostrata. Am J Physiol* 228:420–31.

Schnierla, T.C. (1959). An evolutionary and developmental theory of biphasic processes underlying approach and withdrawal. In: *The Nebraska Symposium on Motivation*, M.R. Jones (ed.), pp. 1–43. Lincoln: University of Nebraska Press.

Schnierla, T.C. (1965). Aspects of stimulation and organization in approach/withdrawal processes underlying vertebrate behavioral development. In: *Advances in the Study of Behavior*, D. Lehrmon, R. Hinde, E. Shaw (eds.). New York: Academic Press.

Schulkin, J. (1981). The appetite for salts in mineral and vitamin deficient rats. *Fed Proc* 40: (abstract).

Schulkin, J. (1982). Behavior of sodium-deficient rats: the search for a salty taste. *J Comp Physiol Psychol* 96:628–34.

Schulkin, J. (1991a). The ingestion of calcium in female and male rats. *Psychobiology* 19:262–4.

Schulkin, J. (1991b). The allure of salt. *Psychobiology* 19:16–21.

Schulkin, J. (1991c). *Sodium Hunger*. Cambridge University Press.

Schulkin, J. (1995). Do reptiles have mineral appetites? Paper presented at the Desert Tortoise Council. Las Vegas, Nevada.

Schulkin, J. (1999). *The Neuroendocrine Regulation of Behavior*. Cambridge University Press.

Schulkin, J., Arnell, P., Stellar, E. (1985). Running to the taste of salt in mineralocorticoid treated rats. *Horm Behav* 19:413–25.

Schulkin, J., McEwen, B.S., Gold, P.W. (1994a). Allostasis, amygdala and anticipatory angst. *Neurosci Biobehav Rev* 18:385–96.

Schulkin, J., Marini, J., Epstein, A.N. (1989). A role for the medial region of the amygdala in mineralocorticoid-induced salt hunger. *Behav Neurosci* 103:178–85.

Schulkin, J., Rozin, P., Stellar, E. (1994b). Curt P. Richter, 1894–1988. In: *Biographical Memoirs*, vol. 65, pp. 311–20. Washington, DC: National Academy Press.

Schwab, E.B., Axelson, M.L. (1984). Dietary changes of pregnant women: compulsions and modifications. *Ecology of Food and Nutrition* 14:143–53.

Scott, E.M., Verney, E.L., Morissey, P.D. (1950). Self-selection of diet. XL. Appetites for calcium, magnesium and potassium. *Am J Nutr* 41:187–201.

Sebert, J.L., Garabedian, M., Chauvenet, M., Maamer, M., Agbomson, F., Brazier, M. (1995). Evaluation of a new solid formulation of calcium and vitamin D in institutionalized elderly subjects: a randomized comparative trial versus separate administration of both constituents. *Rev Rhum Engl Ed* 62:288–94.

Seeley, R.J., Galaverna, O., Schulkin, J., Epstein, A.N., Grill, H.J. (1993). Lesions of the central nucleus of the amygdala. II: Effects on intraoral NaCl intake. *Behav Brain Res* 59:19–25.

Seelig, M.S. (1993). Interrelationship of magnesium and estrogen in cardiovascular and bone disorders, eclampsia, migraine, and premenstrual syndrome. *J Am Coll Nutr* 12:442–58.

Seely, E.W., Brown, E.M., DeMaggio, D.M., Weldon, D.K., Graves, S.W. (1997). A prospective study of calciotropic hormones in pregnancy and post partum: reciprocal changes in serum intact parathyroid hormone and 1,35-dihydoxyvitamin D. *Am J Obstet Gynecol* 176:214–17.

Seely, E.W., Graves, S.W. (1993). Calcium homeostasis in normotensive and hypertensive pregnancy. *Compr Ther* 19:124–8.

Segre, G.V., Brown, E.M. (1996). Secretion, circulating heterogeneity, and metabolism of parathyroid hormone. Chapter 10 in *Primer on the Metabolic Bone Diseases and Disorders of Mineral Metabolism*, 3rd ed., M.J. Favus (ed.). New York: Lippincott-Raven.

Seifert, H., Chesnut, J., DeSouza, E., Rivier, J., Vale, W. (1985). Binding sites for calcitonin gene-related peptide in distinct areas of rat brain. *Brain Res* 346:195–8.

Seki, K., Makimura, N., Mitsui, C., Hirata, J., Nagata, I. (1991). Calcium-regulating hormones and osteocalcin levels during pregnancy: a longitudinal study. *Am J Obstet Gynecol* 164:1248–52.

Selby, P.L., Davies, M., Marks, J.S., Mawer, E.B. (1995). Vitamin D intoxication causes hypercalcaemia by increased bone resorption which responds to pamidronate. *Clin Endocrinol* 43:531–6.

Senior, P.V., Heath, D.A., Beck, F. (1991). Expression of parathyroid hormone-related protein mRNA in the rat before birth: demonstration by hybridization histochemistry. *J Mol Endocrinol* 6:281–90.

Sentipal, J.M., Wardlaw, G.M., Mahan, J., Matkovic, V. (1991). Influence of calcium intake and growth indexes on vertebral bone mineral density in young females. *Am J Clin Nutr* 54:425–8.

Sexton, P.M., Hilton, J.M. (1992). Biologically active salmon calcitonin-like peptide is present in rat brain. *Brain Res* 596:279–84.

Sexton, P.M., McKenzi, J.S., Mendelshon, A.O. (1998). Evidence for a new subclass of calcitonin/calcitonin gene-related peptide binding site in rat brain. *Neurochem Int* 12:323–35.

Sexton, P.M., Schneider, H.G., D'Santos, C.S., Mendelsohn, F.A., Kemp, B.E., Moseley, J.M., Martin, T.J., Findlay, D.M. (1991). Reversible calcitonin binding to solubilized sheep brain binding sites. *Biochem J* 273:179–84.

Shane, E. (1991). Medical management of asymptomatic primary hyperparathyroidism. *J Bone Miner Res [Suppl 2]* 6:S131–4.

Shapiro, M.D., Linas, S.L. (1985). Sodium chloride pica secondary to iron-deficiency anemia. *Am J Kidney Dis* 5:67–8.

Sherman, H.C., Hawley, E. (1992). Calcium and phosphorus metabolism in childhood. *J Biol Chem* 52:375–99.

Sherrington, C. (1906). *The Integrative Action of the Nervous System*. New York: Scribner.

Sherwin, B.B. (1988). Estrogen and/or androgen replacement therapy and cognitive functioning in surgically menopausal women. *Psychoneuroendocrinology* 13:345–57.

Shettleworth, S.J. (1972). Constraints on learning. In: *Advances in the Study of Behavior*, vol. 4, D.S. Lehrman, R.A. Hinde, E. Shaw (eds.), pp. 1–68. New York: Academic Press.

Shetty, P.S., Prentice, A.M., Goldberg, G.R., Murgatroyd, P.R., McKenna, A.P., Stubbs, R.J., Volschenk, P.A. (1994). Alterations in fuel selection and voluntary food intake in response to isoenergetic manipulation of glycogen stores in humans. *Am J Clin Nutr* 60:534–43.

Shiff, E., Friedman, S.A., Sibai, B.M., Kao, L., Schifter, S. (1995). Plasma and placental calcitonin gene-related peptide in pregnancies complicated by severe preeclampsia. *Am J Obstet Gynecol* 173:1405–9.

Shils, M.E. (1969). Experimental human magnesium depletion. *Medicine* 48:61–85.

Shimura, T., Komori, M., Yamamoto, T. (1997). Acute sodium deficiency reduces gustatory responsiveness to NaCl in the parabrachial nucleus of rats. *Neurosci Lett* 236:33–6.

Shulkes, A.A., Covelli, M.D., Denton, D.A., Nelson, J.F. (1972). Hormonal factors influencing salt appetite in lactation. *Aust J Exp Biol Med Sci* 50:819–26.

Silver, J., Naveh-Many, T. (1993). Calcitonin gene regulation in vivo. *Horm Metab Res* 25:470–2.

Silverberg, S.J., Bone, H.G., III, Marriott, T.B., Locker, F.G., Thys-Jacobs, S., Dziem, G., Kaatz, S., Sanguinetti, E.L., Bilezikian, J.P. (1997). Short-term inhibition of parathyroid hormone secretion by a calcium-receptor agonist in patients with primary hyperparathyroidism. *N Engl J Med* 337:1506–10.

Simkiss, K. (1996). Calcium transport across calcium-regulated cells. *Physiol Zool* 69:343–50.

Simkiss, K., Taylor, M.G. (1994). Calcium magnesium phosphate granules: atomistic simulations explaining cell death. *J Exp Biol* 190:131–9.

Simmerly, R.B. (1991). Prodynorphin and proenkephalin gene expression in the anteroventral periventricular nucleus of the rat: sexual differentiation and hormonal regulation. *Mol Cell Neurosci* 2:473–84.

Simmerly, R.B. (1995). Hormonal regulation of limbic and hypothalamic pathways. In: *Neurobiological Effects of Sex Steroids*, P.E. Micevych, R.P. Hammer (eds.). Cambridge University Press.

Simon, H. (1983). *Models of Bounded Rationality*, 2nd ed. Cambridge, MA: MIT Press.

Simoons, F. (1980). Effects of culture: geographical and historical approaches. *Int J Obes* 4:387–94.

Sims, N.R. (1995). Calcium, energy metabolism and the development of selective neuronal loss following short-term cerebral ischemia. *Metab Brain Dis* 10:191–217.

Singer, F.R., Minoofar, P.N. (1995). Bisphosphonates in the treatment of disorders of mineral metabolism. *Adv Endocrinol Metab* 6:259–88.

Skinner, J.D., Pope, J.F., Carruth, B.R. (1998). Alterations in adolescents' sensory taste preferences during and after pregnancy. *J Adolesc Health* 22:43–9.

Skofitsch, G., Jacobowitz, D.M. (1992). Calcitonin- and calcitonin gene-related peptide: receptor binding sites in the central nervous system. Chapter IV in *Handbook of Chemical Neuroanatomy, Vol. 11: Neuropeptide Receptors in the C.N.S.* New York: Academic Press.

Slattery, M.L., Sorenson, A.W., Ford, M.H. (1988). Dietary calcium intake as a mitigation factor in colon cancer. *Am J Epidemiol* 128:504–14.

Smith, D.V., Hanamor, T. (1991). Organization of gustatory sensitivity in hamster. *J Neurosci* 65: 1098–112.

Smith, E.L., Gilligan, C., Smith, P.E., Sempos, C.T. (1989). Calcium supplementation and bone loss in middle-aged women. *Am J Clin Nutr* 50:833–42.

Smith, G.P. (1995). Pavlov and appetite. *Integr Physiol Behav Sci* 30:169–74.

Smith, G.P. (1997). Eating and the American zeitgeist. *Appetite* 29:191–200.

Smith, J.D., Meyer, J.H. (1962). Interaction of dietary sodium and potassium and their influence on energy metabolism. *Am J Physiol* 203:1081–5.

Smogorzewski, M., Islam, A. (1995). Parathyroid hormone stimulates the generation of inositol 1,4,5-triphosphate in brain synaptosomes. *Am J Kidney Dis* 26:814–17.

Snowdon, C.T. (1977). A nutritional basis for lead pica. *Physiol Behav* 18:885–93.

Snowdon, C.T., Sanderson, B.S. (1973). Lead pica produced in rats. *Science* 183:92–4.

Sokoll, L.J., Dawson-Hughes, B. (1992). Calcium supplementation and plasma ferritin concentrations in premenopausal women. *Am J Clin Nutr* 56:1045–8.

Song, Y.H., Ray, K., Liebhaber, S.A., Cooke, N.E. (1998). Vitamin D-binding protein gene transcription is regulated by the relative abundance of hepatocyte nuclear factors 1alpha and 1beta. *J Biol Chem* 273:28408–18.

Sonnenberg, J., Luine, V.N., Krey, L.C., Christakos, S. (1986). 1,25-dihydroxyvitamin $D_3$ treatment results in increased choline acetyltransferase activity in specific brain nuclei. *Endocrinology* 118:1433–9.

Sonnenberg, J., Pansini, A.R., Christakos, S. (1984). Vitamin D-dependent rat renal calcium-binding protein: development of a radioimmunoassay, tissue distribution, and immunologic identification. *Endocrinology* 115:640–8.

Sowers, M., Corton, G., Shapiro, B., Jannausch, M.L., Crutchfield, M., Smith, M.L., Randolph, J.F., Hollis, B. (1993). Changes in bone density with lactation. *JAMA* 269:3130–5.

Sowers, M., Eyre, D., Hollis, B.W., Randolph, J.F., Shapiro, B., Jannausch, M.L., Crutchfield, M. (1995a). Biochemical markers of bone turnover in lactating and nonlactating postpartum women. *J Clin Endocrinol Metab* 80:2210–16.

Sowers, M.F., Hollis, B.W., Shapiro, B., Randolph, J., Janney, C.A., Zhang, D., Schork, A., Crutchfield, M., Stanczy, K.F., Russel-Aulet, M. (1996). Elevated parathyroid hormone-related peptide associated with lactation and bone density loss. *JAMA* 276:549–54.

Sowers, M.F., Randolph, J., Shapiro, B., Jannausch, M. (1995b). A prospective study of bone density and pregnancy after an extended period of lactation with bone loss. *Obstet Gynecol* 85:285–9.

Specker, B.L. (1994). Do North American women need supplemental vitamin D during pregnancy or lactation? *Am J Clin Nutr [Suppl 2]* 59:4905–15.

Specker, B.L. (1996). Evidence for an interaction between calcium intake and physical activity on changes in bone mineral density. *J Bone Miner Res* 11:1539–44.

Specker, B.L., Beck, A., Kalkwarf, H., Ho, M. (1997). Randomized trial of varying mineral intake on total body bone mineral accretion during the first year of life. *Pediatrics* 99(6):12.

Specker, B.L., Lichtenstein, P., Mimouni, F., Gormley, C., Tsang, R.C. (1986). Calcium-regulating hormones and minerals from birth to 18 months of age: a cross-sectional study. II. Effects of sex, race, age, season, and diet on serum minerals, parathyroid hormones, and calcitonin. *Pediatrics* 77:891–6.

Specker, B.L., Tsang, R.C. (1987). Cyclical serum 25-hydroxyvitamin D concentrations paralleling sunshine exposure in exclusively breast-fed infants. *J Pediatr* 110:744–7.

Specker, B.L., Tsang, R.C., Ho, M.L. (1991). Changes in calcium homeostasis over the first year postpartum: effect of lactation and weaning. *Obstet Gynecol* 78:56–62.

Specker, B.L., Tsang, R.C., Ho, M.L., Miller, D. (1987). Effect of vegetarian diet on serum 1,25-dihydroxyvitamin D concentrations during lactation. *Obstet Gynecol* 70:870–4.

Specker, B.L., Tsang, R.C., Hollis, B.W. (1985a). Effect of race and diet on human-milk vitamin D and 25-hydroxyvitamin D. *Am J Dis Child* 139:1134–7.

Specker, B.L., Valanis, B., Hertzberg, V., Edwards, N., Tsand, R.C. (1985b). Sunshine exposure and serum 25-hydroxyvitamin D concentrations in exclusively breast-fed infants. *J Pediatr* 107:372–6.

Specker, B.L., Vieira, N.E., O'Brien, K.O., Ho, M.L., Heubi, J.E., Abrams, S.A., Yergey, A.L. (1994). Calcium kinetics in lactating women with low and high calcium intakes. *Am J Clin Nutr* 59:593–9.

Spector, A.C. (1995). Gustatory function in the parabrachial nuclei: implications from lesion studies in rats. *Rev Neurosci* 6:143–75.

Spector, A.C., Grill, H.J. (1992). Salt taste discrimination after bilateral section of the chorda tympani or glossopharyngeal nerves. *Am J Physiol* 263:R160–76.

Spector, A.C., Schwartz, G.J., Grill, H.J. (1990). Chemospecific deficits in taste detection after selective gustatory deafferentation in rats. *Am J Physiol* 258:R820–6.

Spencer, H. (1872). *The Principles of Psychology*. New York: Appleton-Century-Crofts.

Staal, A., Van Wijnen, A.J., Birkenhager, J.C., Pols, H.A., Prahl, J., DeLuca, H., Gaub, M.P., Lian, J.B., Stein, G.S., Van Leeuwen, J.P., Stein, J.L. (1996). Distinct conformations of vitamin D receptor/retinoid X receptor-alpha heterodimers are specified by dinucleotide differences in the vitamin D-responsive elements of the osteocalcin and osteopontin genes. *Mol Endocrinol* 10:1444–56.

Staaland, H., Jacobsen, E., White, R.G. (1983). The effect of mineral supplements on nutrient concentrations and pool sizes in the alimentary tract of reindeer fed lichens or concentrates during the winter. *Can J Zool* 62:1232–41.

Staaland, H., White, R.G., Luick, J.R., Holleman, D.F. (1980). Dietary influences on sodium and potassium metabolism of reindeer. *Can J Zool* 58:1728–34.

Starling, E.H. (1923). *The Wisdom of the Body*. London: H.K. Lewis & Co.

Steinberg, J., Bindra, D. (1962). Effects of pregnancy and salt-intake on genital licking. *J Comp Physiol Psychol* 55:103–6.

Steiner, J.E. (1977). Facial expressions of the neonate infant indicating the hedonics of food-related chemical stimuli. In: *Taste and Development: The Genesis of Sweet Preferences*, J.M. Weiffenbach (ed.). Bethesda, MD: NIH.

Steiner, J.E. (1979). Human facial expression in response to taste and smell stimuli. *Adv Child Dev Behav* 13:257–95.

Steiner, J.E., Glaser, D., Hawilo, M.E., Berridge, K.C. (in press). To the evolution of facial expressions of positive and negative emotion. *Motivation and Emotion*.

Stellar, E. (1954). The physiology of motivation. *Psychol Rev* 61:5–22.

Stellar, E. (1974). Brain mechanisms in hunger and other hedonic experiences. *Proc Am Philos Soc* 118:276–82.

Stellar, E. (1982). Brain mechanisms in hedonic processes. In: *The Physiological Mechanisms of Motivation*, D.W. Pfaff (ed.), pp. 377–407. New York: Springer-Verlag.

Stellar J.R., Stellar, E. (1985). *The Neurobiology of Motivation and Reward*. New York: Springer-Verlag.

Stepto, R., Keith, L. (1971). Impaired iron balance during pregnancy. *J Natl Med Assoc* 63:87–92.

Sterling, P., Eyer, J. (1988). Allostasis: a new paradigm to explain arousal pathology. In: *Handbook of Life Stress, Cognition and Health*, S. Fisher, J. Reason (eds.). New York: Wiley.

Stern, A.A., Kunz, T.H., Studier, E.H., Oftedal, O.T. (1997). Milk composition and lactational output in the greater spear-nosed bat, *Phyllostomus hastatus*. *J Comp Physiol [B]* 167:389–98.

Stern, J.E., Cardinali, D.P. (1994). Influence of the autonomic nervous system on calcium homeostasis in the rat. *Biol Signals* 3:15–25.

Stern, J.E., Esquifino, A.I., Garcia, B.O., Hacho, M., Cardinali, D.P. (1997). The influence of cervical sympathetic neurons on parathyroid hormone and calcitonin release in the rat: independence of pineal mediation. *J Pineal Res* 22:9–15.

Stevenson, M.F., Rylands, A.B. (1988). The marmosets, genus *Callithrix*. In: *Ecology and Behavior of Neotropical Primates*, vol. 2, R.A. Mittermeier, A.B. Ryland, A. Coimbra-Filho, G.A.B. Fonseca (eds.), pp. 131–222. Washington, DC: World Wildlife Fund.

Stewart, R.E., DeSimone, J.A., Hill, D.L. (1997). New perspectives in gustatory physiology: transduction, development, and plasticity. *Am J Physiol.* 272:C1–26.

Stewart, R.E., Tong, H., McCarty, R., Hill, D.L. (1993). Altered gustatory development in Na(+)-restricted rats is not explained by low Na$^+$ levels in mothers' milk. *Physiol Behav* 53:823–6.

Stiffler, D.F. (1993). Amphibian calcium metabolism. *J Exp Biol* 184:47–61.

Stiffler, D.F. (1996). Exchanges of calcium with the environment and between body compartments in amphibia. *Physiol Zool* 69:418–34.

Strauss, K.I., Jacobowitz, D.M., Schulkin, J. (1994). Dietary calcium deficiency causes a reduction in calretinin mRNA in the substantia nigra compacta–ventral tegmental area of rat brain. *Mol Brain Res* 25:140–2.

Strauss, K.I., Schulkin, J., Jacobowitz, D.M. (1995). Corticosterone effects on rat calretinin in mRNA in discrete brain nuclei and the testes. *Mol Brain Res* 28:81–6.

Strewler, G.J., Nissenson, R.A. (1996). Parathyroid-hormone-related protein. Chapter 12 in *Primer on the Metabolic Bone Diseases and Disorders of Mineral Metabolism*, 3rd ed., M.J. Favus (ed.). New York: Lippincott-Raven.

Stricker, E.M., Sterritt, G.M. (1967). Osmoregulation in the newly hatched domestic chick. *Physiol Behav* 2:117–19.

Struckhoff, G., Turzynski, A. (1995). Demonstration of parathyroid hormone-related protein in meninges and its receptor in astrocytes: evidence for a paracrine meningo-astrocytic loop. *Brain Res* 676:1–9.

Strugnell, S.A., DeLuca, H.F. (1997). The vitamin D receptor – structure and transcriptional activation. *Proc Soc Exp Biol Med* 215:223–8.

Stubbs, R.J., Murgatroyd, P.R., Goldberg, G.R., Prentice, A.M. (1993). Carbohydrate balance and the regulation of day-to-day food intake in humans. *Am J Clin Nutr* 57:897–903.

Studier, E.H., Sevick, S.H., Keeler, J.O., Schenck, R.A. (1994). Nutrient levels in guano from maternity colonies of big brown bats. *J Mammol* 75:71–83.

Stumpf, W.E. (1995). Vitamin D sites and mechanisms of action: a histochemical perspective. Reflections on the utility of autoradiography and cytopharmacology for drug targeting. *Histochem Cell Biol* 104:417–27.

Stumpf, W.E., Hayakawa, N., Koike, N., Hirate, J., Okazaki, A. (1995a). Nuclear receptors for 1,25-dihydroxy-22-oxavitamin D$_3$ (OCT) and 1,25-dihydroxyvitamin D$_3$ in gastric gland neck mucous cells and gastrin enteroendocrine cells. *Histochem Cell Biol* 103:245–50.

Stumpf, W.E., Koike, N., Hayakawa, N., Tokuda, K., Nishimiya, K., Hirate, J., Okazaki, A., Kumaki, K. (1995b). Distribution of 1,25-dihydroxyvitamin D$_3$: in vivo receptor binding in adult and developing skin. *Arch Dermatol Res* 287:294–303.

Stumpf, W.E., O'Brien, L.P. (1987). 1,25(OH)$_2$ vitamin D$_3$ sites of action in the brain, an autoradiographic study. *Histochem Cell Biol* 87:393–406.

Stumpf, W.E., Perez-Delgado, M.M., Li, L., Bidmon, H.J., Tuohimaa, P. (1993). Vitamin D$_3$ (soltriol) nuclear receptors in abdominal scent gland and skin of Siberian hamster (*Phodopus sungorus*) localized by autoradiography and immunohistochemistry. *Histochem Cell Biol* 100:115–19.

Stumpf, W.E., Sar, M., Clark, S.A., DeLuca, H.F. (1982). Brain target sites for 1,25-dihydroxyvitamin D$_3$. *Science* 215:1403–5.

Sumners, C., Gault, T.R., Fregly, M.J. (1991). Potentiation of angiotensin II-induced drinking by glucocorticoids is a specific glucocorticoid type II receptor (GR) mediated event. *Brain Res* 552:283–90.

Sutherland, M.K., Somerville, M.J., Yoong, L.K.K., Bergeron, C., Haussler, M.R., McLachlan, D.R. (1992). Reduction of vitamin D hormone receptor mRNA levels in Alzheimer as compared to Huntington hippocampus: correlation with calbindin-28k mRNA levels. *Mol Brain Res* 13:239–50.

Swanson, L.W., Simmons, D.M. (1989). Differential steroid hormone and neural influences on peptide mRNA levels in CRH cells of the paraventricular nucleus: a hybridization histochemical study in the rat. *J Comp Neurol* 285:413–35.

Sweeny, J.M., Seibert, H.E., Woda, C., Schulkin, J., Haramati, A., Mulroney, S.E. (1998). Evidence for induction of a phosphate appetite in juvenile rats. *Am J Physiol* 275:R1358–65.

Taitano, R.T., Novotny, R., Davis, J.W., Ross, P.D., Wasnich, R.D. (1996). Validity of a food frequency questionnaire for estimating calcium intake among Japanese and white women. *J Am Diet Assoc* 95:804–6.

Talkington, K.M., Gant, N.F., Jr., Scott, D.E., Pritchard, J.A. (1970). Effect of ingestion of starch and some clays on iron absorption. *Am J Obstet Gynecol* 108:262–7.

Tam, C.S., Heersche, J.N.M., Jones, G., Murray, T.M., Rasmussen, H. (1986). The effect of vitamin D on bone in vivo. *Endocrinology* 118(6):2217–24.

Tamura, R., Norgren, R. (1997). Repeated sodium depletion affects gustatory neural responses in the nucleus of the solitary tract of rats. *Am J Physiol* 273:1381–91.

Tannenbaum, G.S., Goltzman, D. (1985). Calcitonin gene-related peptide mimics calcitonin actions in brain on growth hormone release and feeding. *Endocrinology* 116:2685–7.

Tarjan, E., Denton, D.A. (1991). Sodium/water intake of rabbits following administration of hormones of stress. *Brain Res Bull* 26:133–6.

Tarttelin, M.F., Gorski, R.A. (1971). Variations in food and water intake in the normal acyclic female rat. *Physiol Behav* 7:847–52.

Taylor, R., Roper, S. (1994). Ca(2+)-dependent Cl-conductance in taste cells from *Necturus*. *J Neurophysiol* 72:475–8.

Teegarden, D., Lyle, R.M., McCabe, G.P., McCabe, L.D., Proulx, W.R., Michon, K., Knight, A.P., Johnston, C.C., Weaver, C.M. (1998). Dietary calcium, protein, and phosphorus are related to bone density and content in young women. *Am J Clin Nutr* 68:749–54.

Terrado, J., Gerrikagoitia, I., Martinez-Millan, L., Pascual, F., Climent, S., Muniesa, P., Sarasa, M. (1997). Expression of the genes for alpha-type and beta-type calcitonin gene-related peptide during postnatal rat brain development. *Neuroscience* 80:951–70.

Thiels, E., Verbalis, J.G., Stricker, E.M. (1990). Sodium appetite in lactating rats. *Behav Neurosci* 104:742–50.

Thomas, M.K., Lloyd-Jones, D.M., Thadhani, R.I., Shaw, A.C., Deraska, D.J., Kitch, B.T., Vamvakas, E.C., Dick, I.M., Prince, R.L., Finklestein, J.S. (1998). Hypovitaminosis D in medical patients. *N Engl J Med* 338:777–83.

Thorsen, K., Kristoffersson, A., Lorentzon, R. (1995). The effects of brisk walking on markers of bone and calcium metabolism in postmenopausal women. *Calcif Tissue Int* 58:221–5.

Thys-Jacobs, S., Alvir, M.J. (1995). Calcium-regulating hormones across the menstrual cycle: evidence of a secondary hyperparathyroidism in women with PMS. *J Clin Endocrinol Metab* 80:2227–32.

Thys-Jacobs, S., Ceccarelli, S., Bierman, A., Weisman, H., Cohen, M.A., Alvir, J. (1989). Calcium supplementation in premenstrual syndrome. *J Gen Intern Med* 4:183–9.

Thys-Jacobs, S., Silverton, M., Alvir, J., Paddison, P., Rico, M., Goldsmith, S. (1995). Reduced bone mass in women with premenstrual syndrome. *J Women's Health* 4:137–44.

Thys-Jacobs, S., Starkey, P., Bernstein, D., Tian, J. [Premenstrual Syndrome Study Group] (1998). Calcium carbonate and the premenstrual syndrome: effects on premenstrual and menstrual symptoms. *Am J Obstet Gynecol* 179:444–52.

Tilden, C.D., Oftedal, O.T. (1997). Milk composition reflects pattern of maternal care in prosimian primates. *Am J Primatol* 41:195–211.

Tinbergen, N. (1969). *The Study of Instinct*, 2nd ed. Oxford University Press. (Originally published 1951.)

Tolman, E.C. (1949). *Purposive Behavior in Animals and Man*, 2nd ed. Berkeley: University of California Press. (Originally published 1932.)

Tomelleri, R., Grunewald, K.K. (1987). Menstrual cycle and food cravings in young college women. *J Am Diet Assoc* 87:311–15.

Tomoda, S., Kitanaka, T., Ogita, S., Hidaka, A. (1995). Prevention of pregnancy-induced hypertension by calcium dietary supplement: a preliminary report. *J Obstet Gynecol* 21:281–8.

Tordoff, M.G. (1992a). Salt intake of rats fed diets deficient in calcium, iron, magnesium, phosphorus, potassium, or all minerals. *Appetite* 18:29–41.

Tordoff, M.G. (1992b). Influence of dietary calcium on sodium and calcium intake of spontaneously hypertensive rats. *Am J Physiol* 31:R370–81.

Tordoff, M.G. (1994). Voluntary intake of calcium and other minerals by rats. *Am J Physiol* 267:R470–5.

Tordoff, M.G. (1996a). Adrenalectomy decreases NaCl intake of rats fed low-calcium diets. *Am J Physiol* 270:R11–21.

Tordoff, M.G. (1996b). The importance of calcium in the control of salt intake. *Neurosci Biobehav Rev* 20:89–99.

Tordoff, M.G. (1996c). Some basic psychophysics of calcium salt solutions. *Chem Senses* 21:417–24.

Tordoff, M.G. (1997a). NaCl ingestion ameliorates plasma indexes of calcium deficiency. *Am J Physiol* 273:R423–32.

Tordoff, M.G. (1997b). Polyethylene glycol-induced calcium appetite. *Am J Physiol* 273:R587–96.

Tordoff, M.G., Hughes, R.L., Pilchak, D.M. (1993). Independence of salt intake from the hormones regulating calcium homeostasis. *Am J Physiol* 264:R500–12.

Tordoff, M.G., Hughes, R.L., Pilchak, D.M. (1998). Calcium intake by rats: influence of parathyroid hormone, calcitonin, and 1,25-dihydroxyvitamin D. *Am J Physiol* 274:R214–31.

Tordoff, M.G., Okiyama, A. (1996). Daily rhythm of NaCl intake in rats fed low-$Ca^{2+}$ diet: relation to plasma and urinary minerals and hormones. *Am J Physiol* 270:R505–17.

Tordoff, M.G., Rabusa, S.H. (1998). Calcium-deprived rats avoid sweet compounds. *J Nutr* 128:1232–8.

Tordoff, M.G., Ulrich, P.M., Schulkin, J. (1990). Calcium deprivation increases salt intake. *Am J Physiol* 259:R411–19.

Travers, S., Nicklas, K. (1990). Taste bud distribution in the rats. *Anat Rec* 227:372–9.

Trechsel, U., Eisman, J.A., Fischer, J.A., Bonjour, J.P., Fleisch, H. (1980). Calcium-dependent, parathyroid hormone-independent regulation of 1,25-dihydroxyvitamin D. *Am J Physiol* 239:E119–24.

Tsang, R.C., Chen, I., Friedman, M.A., Gigger, M., Steichen, J., Koffler, H., Fenton, L., Brown, D., Pramanik, A., Keenan, W., Strub, R., Joyce, T. (1975). Parathyroid function in infants of diabetic mothers. *J Pediatr* 86:399–404.

Tsang, R.C., Strub, R., Brown, D.R., Steichen, J., Hartman, C., Chen, I.W. (1976). Hypomagnesemia in infants of diabetic mothers: perinatal studies. *J Pediatr* 89:115–19.

Tsuchita, H., Kuwata, T. (1995). Trace lipid from whey-mineral complex enhances calcium availability in young ovariectomized rats. *Br J Nutr* 73:299–309.

Turnlund, J., Margen, S., Briggs, G.M. (1979). Effect of glucocorticoids and calcium intake on bone density and bone, liver and plasma minerals in guinea pigs. *J Nutr* 109:1175–88.

U.S. Department of Health and Human Services. (1991). *Healthy People 2000: National Health Promotion and Disease Prevention Objectives*. DHHS publication (PHS) 91-50212. Washington, DC: U.S. DHHS, Public Health Service.

U.S. Department of Health and Human Services. (1997). Priority area 2 nutrition. In: *Healthy People Review 2000*, pp. 34–6. Hyattsville, MD: DHHS.

Usdin, T.B., Bonner, T.I., Harta, G., Mezey, E. (1996). Distribution of parathyroid hormone-2 receptor messenger ribonucleic acid in rat. *Endocrinology* 137:4285–97.

Utiger, R.D. (1998). The need for more vitamin D. *N Engl J Med* 338:828–9.

Valimaki, M.J., Karkkainen, M., Lamberg-Allardt, C., Laitinen, E., Alhava, E., Heikkinen, J., Impivaara, O., Makela, P., Palmgren, J., Seppanen, R., Vuori, I. (1994). Exercise, smoking, and calcium intake during adolescence and early adulthood as determinants of peak bone mass. *Bone Miner J* 309:230–5.

Van der Maten, G.D. (1995). Low sodium diet in pregnancy: effects on maternal nutritional status. *Eur J Obstet Gynecol Reprod Biol* 61:63–4.

Verhaeghe, J., Bouillon, R. (1992). Calciotropic hormones during reproduction. *J Steroid Biochem Mol Biol* 41:469–77.

Vijande, M., Costales, M., Schiaffini, O., Marin, B. (1978). Angiotensin-induced drinking: sexual differences. *Pharmacol Biochem Behav* 8:753–5.

Villacres, E.C., Wong, S.T., Chavkin, C., Storm, D.R. (1998). Type I adenylyl cyclase mutant mice have impaired mossy fiber long-term potentiation. *J Neurosci* 18:3186–94.

Villar, J., Repke, J.T. (1990). Calcium supplementation during pregnancy may reduce preterm delivery in high-risk populations. *Am J Obstet Gynecol* 163:1124–31.

Villar, J., Repke, J., Belizan, J.M., Pareja, G. (1987). Calcium supplementation reduces blood pressure during pregnancy: results of a randomized controlled clinical trial. *Obstet Gynecol* 70:317–21.

Wade, G.N., Zucker, I. (1969). Taste preferences of female rats: modification by neonatal hormones, food deprivation, and prior experience. *Physiol Behav* 4:935–43.

Wade, G.N., Zucker, I. (1989). Hormonal and developmental influences on rat saccharin preferences. *J Comp Physiol Psychol* 69:291–300.

Walker, A.R., Walker, B.F., Jones, J., Verardi, M., Walker, C. (1985). Nausea and vomiting and dietary cravings and aversions during pregnancy in South African women. *Br J Obstet Gynaecol* 92:484–9.

Walker, J., Ball, M. (1993). Increasing calcium intake in women on a low-fat diet. *Eur J Clin Nutr* 47:718–23.

Walker, R.M., Linkswiler, H.M. (1972). Calcium retention in the adult human male as affected by protein intake. *J Nutr* 102:1297–302.

Walters, M.R., Fischette, C.T., Fetzer, C., May, B., Riggle, P.C., Tibaldo-Bongiorno, M., Christakos, S. (1992a). Specific 1,25-dihydroxyvitamin D binding sites in choroid plexus. *Eur J Pharmacol* 213:309–11.

Walters, M.R., Hunziker, W., Norman, A.W. (1980). Unoccupied 1,25-dihydroxyvitamin D receptors. Nuclear/cytosol ratio depends on ionic strength. *J Biol Chem* 255:6799–805.

Walters, M.R., Kollenkirchen, U., Fox, J. (1992b). What is vitamin D deficiency? *Proc Soc Exp Biol Med* 199:385–93.

Wardlaw, G.M., Pike, A.M. (1986). The effect of lactation on peak adult shaft and ultra-distal forearm bone mass in women. *Am J Clin Nutr* 44:283–6.

Warnock, G.M., Duckworth, J. (1994). Changes in the skeleton during gestation and lactation in the rat. *Biochem J* 38:220–4.

Watney, P.J., Rudd, B.T. (1974). Calcium metabolism in pregnancy and in the newborn. *J Obstet Gynaecol Br Commonw* 81:210–19.

Watson, J.B., Lee, K., Klein, R., Klein, B.E., Koch, D.D. (1997). Epidemiological evidence for the disruption of ionized calcium homeostasis in the elderly. *J Clin Epidemiol* 50:845–9.

Watts, A.G., Sanchez-Watts, G. (1995). Region-specific regulation of neuropeptide mRNAs in rat limbic forebrain neurones by aldosterone and corticosterone. *J Physiol* (*Lond*) 484:721–36.

Weaver, C.M. (1994). Age related calcium requirements due to changes in absorption and utilization. *J Nutr* 124:1418S–25S.

Weaver, C.M., Martin, B.R., Plawecki, K.L., Peacock, M., Wood, O.B., Smith, D.L., Wastney, M.E. (1995a). Differences in calcium metabolism between adolescent and adult females. *Am J Clin Nutr* 61:577–81.

Weaver, C.M., Peacock, M., Martin, B.R., Plawecki, K.L., McCabe, G.P. (1996). Calcium retention estimated from indicators of skeletal status in adolescent girls and young women. *Am J Clin Nutr* 64:67–70.

Weaver, D.R., Deeds, J.D., Lee, K., Segre, G.V. (1995b). Localization of parathyroid hormone-related peptide (PTHrP) and PTH/PTHrP receptor mRNAs in rat brain. *Brain Res Mol Brain Res* 28:296–310.

Webb, A.R., DeCosta, B.R., Holick, M.F. (1989). Sunlight regulates the cutaneous production of vitamin $D_3$ by causing its photodegradation. *J Clin Endocrinol Metab* 68:882–7.

Webb, A.R., Kline, L., Holick, M.F. (1998). Influence of season and latitude on the cutaneous synthesis of vitamin $D_3$: exposure to winter sunlight in Boston and Edmonton will not promote vitamin $D_3$ synthesis in human skin. *J Clin Endocrinol Metab* 67:373–8.

Webb, A.R., Pilbeam, C., Hanafin, N., Holick, M.F. (1990). An evaluation of the relative contributions of exposure to sunlight and of diet to the circulating concentrations of 25-hydroxyvitamin D in an elderly nursing home population in Boston. *Am J Clin Nutr* 51:1075–81.

Weingarten, H.P., Elston, D. (1990). The phenomenology of food cravings. *Appetite* 15:231–46.

Weir, E.C., Brines, M.L., Ikeda, K., Burtis, W.J., Broadus, R.J. (1990). Parathyroid hormone-related peptide gene is expressed in the mammalian central nervous system. *Proc Natl Acad Sci USA* 87:108–12.

Weir, J.S. (1969). Chemical properties and occurrence on Kalahari sand of salt licks created by elephants. *J Zool London* 158:293–310.

Weisinger, J.R., Bellorin-Font, E. (1998). Magnesium and phosphorus. *Lancet* 352:391–6.

Weisinger, R.S., Blair-West, J.R., Denton, D.A., Tarjan, E. (1997). Role of brain angiotensin II in thirst and sodium appetite of sheep. *Am J Physiol* 273:187–96.

Weisinger, R.S., Denton, D.A., McKinley, M.J., Nelson, J.F. (1978). ACTH induced sodium appetite in the rat. *Pharmacol Biochem Behav* 8:339–42.

Welch, J.G., Reese, W.H., Smith, A.M. (1973). Calcium supplement consumption during calcium deficiency. *J Dairy Sci* 56:1385 (abstract).

Wheatly, M.G. (1990). Postmolt electrolyte regulation in crayfish. Calcium budget, hemolymph ions and tissue calcium ATPase. *Am Zool* 30:63a.

Wheatly, M.G. (1996). An overview of calcium balance in crustaceans. *Physiol Zool* 69:351–82.

Wheatly, M.G., Greenaway, P. (1996). Calcium regulation: mechanisms and control in crustaceans and lower vertebrates – an introduction. *Physiol Zool* 69:340–2.

Whitfield, G.K., Hsieh, J.C., Nakajima, S., MacDonald, P.N., Thompson, P.D., Jurutka, P.W., Haussler, C.A., Haussler, M.R. (1995). A highly conserved region in the hormone-binding domain of the human vitamin D receptor contains residues vital for heterodimerization with retinoid X receptor and for transcriptional activation. *Mol Endocrinol* 9:1166–79.

Whitlock, E.P. (1988). *The Calcium Plus Workbook*. New Canaan, CT: Keat Publishing.

Whitlock, R.H., Kesslee, M.J., Tasker, J.B. (1975). Salt (sodium) deficiency in dairy cattle: polyuria and polydipsia as prominent clinical features. *Cornell Vet* 65:512–26.

Widmark, E.M. (1944). The selection of food. III. Calcium. *Acta Physiol Scand* 322–8.

Wijewardene, K., Fonseka, P., Goonaratne, C. (1994). Dietary cravings and aversions during pregnancy. *Indian J Public Health* 3:95–100.

Wiles, G.J., Weeks, H.P. (1986). Movements and use patterns of white-tailed deer *Odocoileus virginianus* visiting natural licks. *J Wild Manag* 50:487–96.

Wilkins, L., Richter, C. (1940). A great craving for salt by a child with cortico-adrenal insufficiency. *JAMA* 114:866–8.

Willett, W. (1990). *Nutritional Epidemiology*. Oxford University Press.

Wilson, J.F. (1987). Severe reduction in food intake by pregnant rats resembles a learned food aversion. *Physiol Behav* 41:291–5.

Wilson, P., Horwath, C. (1996). Validation of a short food frequency questionnaire for assessment of dietary calcium intake in women. *Eur J Clin Nutr* 50:220–8.

Wilson, S.G., Retallack, R.W., Kent, J.C., Worth, G.K., Gutteridge, D.H. (1990). Serum free 1,25-dihydroxyvitamin D and the free 1,25-dihydroxyvitamin D index during a longitudinal study of human pregnancy and lactation. *Clin Endocrinol (Oxf)* 32:613–22.

Wimalawansa, S.J., el-Kholy, A.A. (1993). Comparative study of distribution and biochemical characterization of brain calcitonin gene-related peptide receptors in five different species. *Neuroscience* 54:513–19.

Wimalawansa, S.J., Supowit, S.C., DiPette, D.J. (1995). Mechanism of the antihypertensive effects of dietary calcium and role of calcitonin gene related peptide in hypertension. *Can J Physiol Pharmacol* 73:981–5.

Winsky, L., Montpied, P., Arai, R., Martin, B.M., Jacobowitz, D.M. (1992). Calretinin distribution in the thalamus of the rat: immunohistochemical and in situ hybridization histochemical analyses. *Neuroscience* 50:181–96.

Winsky, L., Nakata, H., Martin, B.M., Jacobowitz, D.M. (1989). Isolation, partial amino acid sequence, and immunohistochemical localization of a brain-specific calcium-binding protein. *Proc Natl Acad Sci USA* 86:10139–43.

Wirsig, C.R., Grill, H.J. (1982). Contribution of the rat's neocortex to ingestive control: I. latent learning for the taste of NaCl. *J Comp Physiol Psychol* 96:615–27.

Wojcicka-Jagodzinska, J., Romejko, E., Piekarski, P., Czajkowski, K., Smolarczyk, R., Lipinski, T. (1998). Second trimester calcium-phosphorus-magnesium homeostasis in women with threatened preterm delivery. *Int J Gynaecol Obstet* 61:121–5.

Wojcik, S.F., Schanbacher, F.L., McCauley, L.K., Zhou, H., Kartsogiannis, V., Capen, C.C., Rosol, T.J. (1998). Cloning of bovine parathyroid hormone-related protein (PTHrP) cDNA and expression of PTHrP mRNA in the bovine mammary gland. *J Mol Endocrinol* 20:271–80.

Wolf, G. (1969a). Innate mechanisms for regulation of sodium intake. In: *Olfaction and Taste*, C. Pfaffmann (ed.). New York: Rockefeller University Press.

Wolf, G. (1969b). Effects of mineralocorticoid antagonists on sodium appetite. In: *VIII International Congress of Nutrition*, Prague.

Wolf, G. (1982). Refined salt appetite methodology for rats demonstrated by assessing sex differences. *J Comp Physiol Psychol* 96:1016–21.

Wolf, G., Schulkin, J. (1980). Brain lesions and sodium appetite. In: *Biological and Behavioral Aspects of Salt Intake*, M. Kare (ed.). New York: Academic Press.

Wood, R.J., Fleet, J.C., Cashman, K., Bruns, M.E., DeLuca, H.F. (1998). Intestinal calcium absorption in the aged rat: evidence of intestinal resistance to 1,25(OH)$_2$ vitamin D. *Endocrinology* 139:3843–8.

Wood-Gush, D.G., Kare, M.R. (1966). The behaviour of calcium-deficient chickens. *Br Poult Sci* 7:285–90.

Woodhead, J.S. (1990). The measurement of circulatung parathyroid hormone. *Clin Biochem* 23:17–21.

Woods, S.C., Vasselli, J.R., Milam, K.M. (1977). Iron appetite and latent learning in rats. *Physiol Behav* 29:623–6.

Woods, S.C., Weisinger, R.S. (1970). Pagophagia in the albino rat. *Science* 169:1334–6.

Woodside, B., Millelire, L. (1987). Self-selection of calcium during pregnancy and lactation in rats. *Physiol Behav* 39:291–5.

Worthington-Roberts, B., Little, R.E., Lambert, M.D., Wu, R. (1989). Dietary cravings and aversions in the postpartum period. *J Am Diet Assoc* 89:647–51.

Wright, P., Crow, R.A. (1973). Menstrual cycle: effect on sweetness preferences in women. *Horm Behav* 4:387–91.

Wyewardene, K., Fonseka, P., Goonaratne, C. (1994). Dietary cravings and aversions during pregnancy. *Indian J Public Health* 38:95–8.

Yamin, M., Gorn, A.H., Flannery, M.R., Jenkins, N.A., Gilbert, D.J., Copeland, N.G., Tapp, D.R., Krane, S.M., Goldring, S.R. (1994). Cloning and characterization of a mouse brain calcitonin receptor complementary deoxyribonucleic acid and mapping of the calcitonin receptor gene. *Endocrinology* 135:2635–43.

Yan, L., Prentice, A., Dibba, B., Jarjou, L.M., Sterling, D.M., Fairweather-Tait, S. (1996). The effect of long-term calcium supplementation on indices of iron, zinc, and magnesium status in lactating Gambian women. *Br J Nutr* 76:821–31.

Yasui, M., Ota, K., Garruto, R.M. (1995). Effects of calcium-deficient diets on manganese deposition in the central nervous system and bones of rats. *Neurotoxicology* 16:511–17.

Yasui, M., Yase, Y., Ota, K., Garruto, R.M. (1991). Evaluation of magnesium, calcium and aluminum metabolism in rats and monkeys maintained on calcium-deficient diets. *Neurotoxicology* 12:603–14.

Yensen, R. (1959a). Some factors affecting taste sensitivity in man. I: Food intake and time of day. *Q J Exp Psychol* 11:221–9.

Yensen, R. (1959b). Some factors affecting taste sensitivity in man. III: Water deprivation. *Q J Exp Psychol* 11:239–48.

Yetgin, S., Hincal, F., Basaran, N., Ciliv, G. (1992). Serum selenium status in children with iron deficiency anemia. *Acta Haematol* 88:185–8.

Young, P.T. (1948). Appetite, palatability and feeding habit: a critical review. *Psychol Bull* 45:289–320.

Young, P.T., Chaplin, J.P. (1949). Studies of food preference, appetite, and dietary habit. In: *Comparative Psychology Monographs*, vol. 19, no. 5. Berkeley: University of California Press.

Young, P.T., Falk, J.L., Kappauf, W.E. (1956). Running activity and preference as related to concentration of sodium-chloride solutions. *J Comp Physiol Psychol* 49:569–75.

Yu, S., Rogers, Q.R., Morris, J.G. (1997). Absence of a salt (NaCl) preference or appetite in sodium-replete or depleted kittens. *Appetite* 29:1–10.

Zhao, X.P., Wang, S., Xia, Y.H. (1996). Effects of calcitonin injected into various brain areas on pain threshold and $Ca^{2+}$ in rats. *Chung Kuo Yau Li Hsueh Pao* 17:218–20.

Zierold, C., Darwish, H.M., DeLuca, H.F. (1994). Identification of a vitamin D-response element in the rat calcidiol (25-hydroxyvitamin $D_3$) 24-hydroxylase gene. *Proc Natl Acad Sci USA* 91:900–2.

Zinaman, M.J., Hickey, M., Tomai, T.P., Albertson, B.D., Simon, J.A. (1990). Calcium metabolism in postpartum lactation: the effect of estrogen status. *Fertil Steril* 54:465–9.

Zucker, I. (1965). Short-term salt preference of potassium-deprived rats. *Am J Physiol* 208(6):1071–4.

Zucker, I. (1969). Hormonal determinants of sex differences in saccharin preference, food intake and body weight. *Physiol Behav* 4:595–602.

# Index

sodium: calcium deficiency and ingestion of, 45–9, *50f*, 57, 105; and calcium retention, 126; estrogen levels and extracellular-fluid regulation of, 62–3; gonadal steroid hormones and ingestion of, 59–61; hormones and ingestion of calcium and, 105–6; and ingestive behavior as adaptive response, 14–15; ingestive behavior and gustatory system, 28; and innate specific appetite, 18; lactation and intake of, 66; plants and calcium sensors, 143n4; pregnancy and ingestion of, 65, 66
South Africa, study of ethnicity and food cravings during pregnancy, 79
speciation and variation, and evolution, 11
Sri Lanka, and food cravings during pregnancy, 79
stress hormones: calcium deficiency and salt appetite in rats, 105; and calcium ingestion, 103, 105
stria terminalis, bed nucleus of, 108
sunlight, vitamin D and exposure to, 112, 113–14
supplements, dietary: and calcium, 110, 118–19, 121, 127, 130, 141; and vitamin D, 113, 124
sweets: menstrual cycle and craving for, 64; as sources of calcium, *7f*, 141

taste: and calcium, 20, 30, 32–3; and facial expressions of infants, 19–20; four groups of, 28–9; functions of, 27–8; gender and reactions to sodium, 60; and taste-aversion learning, 17
taste buds, 28
thalamic gustatory region, lesions of, 37, *38f*
thiamine: deficiency of and learned appetite, 18; and ingestive behavior, 17
tissues, and distribution of calcium in human, *4t*
tongue, and gustatory anatomy, 28, *29*

Tordoff, M. G., 47, 91, 105
transport, of calcium: calcium-binding proteins and calcium appetite, 102; and calcium homeostasis, 143n1; and cellular mechanisms, 108; and evolution, 3; from mother to fetus during pregnancy, 72–3

ultraviolet B radiation, 114
urbanization, and calcium or vitamin D deficiencies in diet, 112

vegetables, as sources of calcium, *7f*, 140
vegetarian diets, and calcium ingestion, 126
vertebrates: amount of calcium in body of, 2; and evolution of calcium regulation, 3
vitamin D: and bone health, 123–4; and brain, 91–2; and calcium in diet, 112–15; and calcium homeostasis, 85–6; and calcium ingestion, 67–70, 92–6, 123–4; and calcium regulation in iguanas, 11, *12f*; and colorectal cancer, 132–3; deficiencies of, 67–70, 95–6; and gene expression, *87f*; and hypocalcemia, 129; increased concentrations of during pregnancy, 73; and preeclampsia, 130

wanting, and palatability judgments, 20
water: and concentration of calcium in ocean, 2, 24; estrogen levels and extracellular-fluid regulation of sodium and, 62–3; pregnancy and ingestion of, 70
women's health, and importance of calcium, 5, 138. *See also* lactation; pregnancy
World War II, and healthy diets, 23

Zaire, and calcium intake during pregnancy and lactation, 82
zinc, and pica, 54

Printed in the United States
By Bookmasters